SpringerBriefs in Energy
Energy Analysis

Series Editor: Charles A.S. Hall

rlarling Sophia :
This is a revisiting of
the 1972 MIT
paper —
 Enjoy —
 I love you, toots.
WASABI — Marco

D1452589

For further volumes:
http://www.springer.com/series/10041

Ugo Bardi

The Limits to Growth Revisited

Foreword by Ian Johnson

 Springer

Ugo Bardi
Università di Firenze
Dipartimento di Chimica
Polo Scientifico di Sesto
Fiorentino
50019 Firenze
Italy
ugo.bardi@unifi.it

ISSN 2191-5520   e-ISSN 2191-5539
ISSN 2191-7876
ISBN 978-1-4419-9415-8   e-ISBN 978-1-4419-9416-5
DOI 10.1007/978-1-4419-9416-5
Springer New York Dordrecht Heidelberg London

Library of Congress Control Number: 2011928230

Springer is part of Springer Science+Business Media (www.springer.com)

This book is dedicated to the memory of Matthew Simmons (1943–2010), supporter of "peak oil" studies, advocate of renewable energy, and one of the first to take a new look at "The Limits to Growth" in the twenty-first century.

From "Agamennon" by Aeschilus; words of Cassandra (*)

More bright shall blow the wind of prophecy,
And as against the low bright line of dawn
Heaves high and higher yet the rolling wave,
So in the clearing skies of prescience
Dawns on my soul a further, deadlier woe,
And I will speak, but in dark speech no more.

Foreword

The report to the Club of Rome entitled *The Limits to Growth* (LTG) was published in 1972 and is regarded as one of the most influential books of the twentieth century. Its messages left deep impressions on many individuals worldwide who today can be found in influential positions in politics, administration, civil society, or academia. LTG was the catalyst that opened our minds to the truth about the precarious state of the planet. Its authors founded a body of thinking that continues to this day.

The core message that was contained in LTG, and reinforced in two later versions, most recently in 2004, was that in a finite world, material consumption and pollution cannot continue to grow forever. When LTG was published, it created a widespread discussion which quickly turned from a scientific debate into a political one, driven by ideology and special interests. Despite often being depicted as such, none of the LTG studies were a prediction of unavoidable doom. Rather, LTG "pleaded for profound proactive, societal innovation through technological, cultural and institutional change, in order to avoid an increase in the ecological footprint of humanity beyond the carrying capacity of the planet." Although LTG warned about the direction the world was heading in and the possibility of collapse, it was also optimistic about changes that could be brought about if society began to recognize the state of global development and take corrective actions. All too often critics chose to ignore the core messages of the report.

What was considered as futuristic 40 years ago has now become the reality of today. We are already in "overshoot" in a number of fields, and it is becoming obvious to more and more people that we have entered into a dangerous era: global warming, peak oil, biodiversity extinction, and reduced ecosystem resilience to name a few. In some cases, we have substitutes but at very high costs; in other cases such as biodiversity, we have much less knowledge on the impacts of the destruction taking place.

As the Secretary General of the Club of Rome, I welcome the range of issues covered in this timely book, *The Limits to Growth Revisited* by Ugo Bardi. His book allows us to better understand the controversy following the release of LTG while at the same time gain insights into "dynamic modeling" and some of the key arguments debated. It should be a "must read" for those interested in our common future, and a better understanding of how good science and analysis can be misinterpreted and maligned.

Ian Johnson

Secretary General of the Club of Rome
and former Vice President for Sustainable
Development, World Bank

Acknowledgements

Let me thank first of all Jay Wright Forrester, pioneer of system dynamics; though in his late 80s when I started writing this book, he nevertheless, provided me with insights, information, and much help on the story of world dynamics. I would also like to thank Dennis Meadows and Jorgen Randers, two of the authors of the first "The Limits to Growth" book who helped me delve into the complex story of their study. I would also like to thank Karl Wagner and Ian Johnson the Club of Rome for their support as well as Colin Campbell and all the members of the Association for the Study of Peak Oil (ASPO).

Finally, I wish to thank all those who offered advice, criticism, or who helped in other ways in the making of this book: Janet Barocco, Nicola dall'Olio, Charles Hall, Richard Heinberg, Marek Kolodzej, Magne Myrtveit, Giorgio Nebbia, David Neiman, Jorgen Noorgard, William Nordhaus, Alessandro Lavacchi, David Packer, Douglas Reynolds, David Roser, Gail Tilverberg, Leigh Yaxley, and Richard Worthington.

Contents

Chapter 1
Introduction

The twentieth century was a period of unprecedented economic expansion, population growth, and technological progress. All that generated great optimism in the future of humankind, an attitude that was especially prevalent in the years after the Second World War. The economic growth experienced up to then was seen as the natural state of things, while population growth was not perceived as a problem. Space travel, flying cars, extension of human lifespan, energy "too cheap to meter," and many other technological wonders were seen as around the corner.

Yet, the twentieth century was also the time when fundamental questions about the destiny of humankind were asked. How long could the economy keep growing on a finite planet? Could population growth continue for a long time at the rates experienced up to then? What were the ultimate material boundaries of civilization?

In the late 1960s, a think tank of intellectuals who referred to themselves as the "Club of Rome" started seeking quantitative answers to these questions. For their quest, they sought the help of a group of scientists at the Massachusetts Institute of Technology in Cambridge, MA, who engaged in the ambitious task of modeling the evolution of the worldwide economic system over a time span of more than a century. The result was the study published in 1972 under the title of "The Limits to Growth" (LTG) (Meadows et al. 1972) with the subtitle "A Report to the Club of Rome."

The results of the study were far from optimistic. The simulations showed that, in a "business as usual" set of assumptions, economic growth could not be maintained throughout the twenty-first century. In the model, the gradual depletion of nonrenewable resources, coupled with increasing pollution and population growth, resulted in the peaking and the subsequent decline of the world's industrial and agricultural production. That was followed, later on, by the decline of population as a consequence of the reduced availability of food and services.

The "base case" scenario of the study, the model that used parameters based on the best, generated the start of the collapse of industrial and agricultural production sometime during the second decade of the twenty-first century. Scenarios based on more optimistic assumptions on the availability of natural resources could only postpone the collapse, but not avoid it. Only specific interventions to curb economic growth

U. Bardi, *The Limits to Growth Revisited*, SpringerBriefs in Energy: Energy Analysis,
DOI 10.1007/978-1-4419-9416-5_1, © Ugo Bardi 2011

and reduce the consumption of natural resources could put humankind on a sustainable path and avoid collapse.

Arriving in a world that had experienced rapid economic growth for at least two decades, these results were a shock. The book generated enormous interest and the debate on its validity raged. Some recognized it as a milestone in human thought, while others dismissed it as a worthless exercise based on flawed assumptions.

Although the debate was initially balanced, gradually the critics gained ground and, eventually, the negative opinion prevailed. By the 1990s, the LTG study and its authors were literally submerged by a wave of criticism, often based on the claim that the study had made wrong predictions in estimating that some important mineral resources would run out before the end of the twentieth century (Bailey 1989).

The negative opinion about LTG appears to be still prevalent today. However, there are also evident signs of a reappraisal in progress. The new trend started with the work of Matthew Simmons (2000) who reviewed the 1972 LTG book and found it still valid in describing the world's situation. Several other authors favorably re-examined the study. Myrtveit (2005) found that the debate was far from being settled against the LTG case. Turner (2008) and Hall and Day (2009) found that the "base case" scenario of LTG has described remarkably well the behavior of parameters such as population growth, resource depletion, and pollution during the past 30 years. Bardi (2008a) revisited some of the common misperceptions about the results of the study, observing that the widely held opinion that LTG had made "wrong predictions" was based on a flawed reading of some of the data reported in the 1972 book. Bardi (2008b) and Hall and Day (2009) also noted the similarity of the LTG models with those developed by Hubbert (1956) for modeling crude oil production. Other authors (Nebbia 2001; Gardner 2004; Busby 2006; Nørgård et al. 2010; Heinberg 2010) pointed out that LTG may have been dismissed on the basis of wrong arguments and stressed the need of a reappraisal. Some of the authors of the first LTG have also published a new study based on the same methods, confirming the results of the 1972 study (Meadows et al. 2004).

On the other hand, the LTG study continues to receive negative criticism, even in the recent scientific literature (see, e.g., Popper et al. 2005; Radetzki 2010). It seems, therefore, that the debate that had raged in the 1970s is flaring again with two opposite lines of thought facing each other. One, favorable to LTG, emphasizes the physical limits of the planet we live on; the other, unfavorable, stresses the importance of progress and human ingenuity in dealing with the challenges ahead.

It has been said – correctly – that prediction is very difficult, especially about the future. But, even though we cannot predict the future, the worst mistake we can make is to ignore it. The future has the nasty habit of transforming itself into the present when it catches people by surprise with all sorts of unpleasant events from financial crashes to giant hurricanes and tsunamis. Sometimes we call these events "acts of God," but it is not God's fault if we are so often caught unprepared. What we need in order to deal with the future is not exact predictions. Warnings are enough.

In our times, we have excellent predictive tools in the form of science-based models. Although the future remains impossible to predict with certainty, these models can still provide us with useful warnings. Unfortunately, we often choose to ignore warnings when the consequences of heeding them are unpleasant. It is, in the end, the destiny of Cassandra who tried, in vain, to warn her fellow Trojan citizens about the impending danger. Today, we often use the term "Cassandra" as an insult, forgetting that Cassandra had been right. We could call this tendency of ignoring warnings the "Cassandra effect."

The warnings that we received in 1972 from LTG are becoming increasingly more worrisome as reality seems to be following closely the curves that the LTG scenarios had generated. Several recent events can be seen as ominous hints that the collapse of the world's economy foreseen by most LTG scenarios may be around the corner. One of these events is the run of oil prices that reached a peak of almost 150 dollars per barrel in 2008. This rapid rise led to renewed fears of exhaustion of a crucial resource for industrial society. Indeed, crude oil may have reached a world-wide maximum level of production ("peak oil") in 2008 (Aleklett et al. 2010), as had been predicted some years before (Campbell and Laherrere 1998). Then, the Great Recession and the crash of the world's financial markets in 2008–2009 brought about the growing belief that deficit spending is no longer an appropriate remedy because of the uncertainty of paying the bill with future economic growth.

If what we are seeing is a first signal of general collapse of the economy, by ignoring the warning that we received in 1972 we lost more than 30 years that could have been used to prevent it. But, it may not be too late to avoid a disastrous col-lapse, provided that we make the right choices and that we use scientific tools in order to understand what the future has in store for us. It may be time to revisit the LTG story and to learn from our previous mistakes.

The authors of the third book of the LTG series (Meadows et al. 2004) state that their book "is a model of what is in our minds, and creating it has altered what we know" (p. 121). These words are valid also for describing the present book, which can only be a model of what is in the mind of the author, including the fact that it has altered what the author knows. Taking this limitation into account, *The Limits to Growth Revisited* is an attempt to present the whole story of LTG in its various sci-entific and historical aspects. The book describes how models were used for the study as well as the aims of the authors in using them. It reviews the debate that took place after publication by examining the objections and the criticism that were made against it, especially by economists. Two points are examined in detail as they are crucial to the validity of the LTG approach: (1) Are mineral resources really finite? (2) Can technology save us from collapse and, in particular, from mineral depletion? Subsequently, *The Limits to Growth Revisited* describes how LTG was rejected and demonized and how it is now undergoing a reappraisal that is highlighting how the early rejection was based on a flawed reading of the text of the study and on a mis-understanding of its purposes. Finally, the book provides a discussion on the rele-vance of the study today and what can we still learn from it.

This book is not aimed at the specialist in system dynamics and tries to present a view of the subject that is based on relatively simple, "mind-sized" concepts. Examples are drawn from the work of the author on modeling economic systems such as the pre-petroleum economy of whaling in the nineteenth century and oil extraction in the twentieth century.

Chapter 2
The Story of "The Limits to Growth"

The end of the Second World War brought a period of great prosperity for the Western World. It was the time of suburban housing, of two cars for every family, of a refrigerator in the kitchen, and of air travel which was not any more a privilege for the rich. It was the time of plastics, of antibiotics, of television, and of the first computers. Those were also the years of the start of the exploration of space. The first Sputnik satellite was launched in 1957. Only 12 years later, in 1969, a man set foot for the first time on the Moon.

All this was taking place together with a rapid increase in the use of energy and of raw materials. In the late 1960s, crude oil took the lead from coal as the most important energy source in the world. Cheap fuels obtained from crude oil generated the urban reality that we see today in most of the Western World: large suburban areas inhabited by commuters who use their cars to reach their workplaces in urban centers.

Oil was not just a source of fuels for private cars. Only with the cheap energy that came mostly from oil was it possible to generate the flux of mineral commodities that was indispensable for modern industry. Plastics become an everyday commodity, replacing wood and natural textiles. Steel, once an expensive hallmark of military power, became cheap and commonplace also for civilian uses. Aluminum became a standard material for all kinds of applications, from cooking to the aeronautics industry. The growing hi-tech industry could find exotic and rare elements at reasonable prices for the applications in electronics, aerospace, chemistry, and engineering. Finally, the "green revolution" in agriculture was the result of a combination of mechanization and artificial fertilizers; both were made possible by cheap crude oil and natural gas. It was the great increase in agricultural productivity generated by the green revolution that made it possible to feed the growing world population; a trend that is continuing up to now.

It was known that crude oil – and fossil fuels in general – were a finite resource, but that did not affect the general optimism of the overwhelming majority of scientists, engineers, economists, and other thinkers. Already in the 1950s, new forms of energy were believed to be ready to usher in an even greater abundance than what fossil fuels had brought. The first nuclear reactor designed for the production of electric power went into operation in the US in 1951. It was the start of what was seen as an onrushing wave of reactors that, eventually, would provide humankind with a new era of prosperity; greater than anything experienced before. The finiteness of mineral uranium resources was recognized, but it was believed that it could be overcome by developing "breeder" reactors and, eventually, by fusion reactors that would provide us with energy "too cheap to meter" (Strauss 1954).

With so much prosperity and with growth seeming unstoppable, some outlandish ideas did not seem far-fetched. In the 1950s, people seriously thought of flying cars, of one-use clothes, of atomic planes, and of weekends on the Moon for the whole family. And why not colonizing other planets? What prevented us from exploring the whole Galaxy? The rise of technological progress had convinced most people that human ingenuity could overcome all problems that came from limited resources.

And yet, in the middle of so much optimism, a new consciousness was appearing. If the Western World was experiencing so much prosperity, it was also easy to see that the rest of the world was being left behind. Those regions optimistically defined as "developing countries" were not developing at all and that was true even for parts of the affluent countries (Harrington 1962). True, the green revolution had reduced the frequency of famines in the world but, nevertheless, something prevented the Western prosperity from spilling over to people who could not even dream of two cars per family or of flying to Hawaii for their vacations.

Was it just a question of time? Perhaps, just waiting long enough, the magic of the free market would accomplish the miracle of putting everyone on the development path. Or was it a cultural problem? Perhaps, by diffusing Western ideas, Western culture, and the Western way of life, including hamburger and hot dogs, everyone in the world would learn how to set up a business and be ready for the local industrial revolution.

But, perhaps, the problem was different. Perhaps there just was something wrong in the idea that the earth had sufficient natural resources to provide everyone with the same way of life that had become standard in the Western World. So, it started to appear clear to some that there were limits to human growth on a finite planet.

The concept of limits to growth was nothing new. Even in ancient times, periods of crisis had led people to wonder about what we call today "overpopulation." In modern times, the problem was studied by Thomas Robert Malthus, who published his "An essay on the principle of population" from 1798 to 1826; a work that remains today a milestone in the understanding of the physical limits that humankind faces. But Malthus was misunderstood already in his times, with Thomas Carlyle defining economics as "the dismal science" with a reference to Malthus's work. In later times, Malthus was accused of all sorts of errors, in particular of having predicted a catastrophe that did not occur – a common destiny for all those who predict a not so bright future. But Malthus was not "merely a foreteller of evil to come" (McCleary 1953). Rather, he was the first to understand that biological populations tend to grow until they have reached the limit that their environment can support. He had a profound influence in biology, for instance, where he influenced Darwin's thought, and in economics as well. The classical economists of the eighteenth and nineteenth century, Adam Smith, Ricardo, Mills, Jevons, and others, had a clear view of the limits of the economic system and adapted their thought accordingly.

The vision of the classical economists became outmoded with the great rush of optimism that started in the second half of the twentieth century. But, even in the midst of what was perhaps the fastest paced phase of economic growth of human history, some people were asking themselves a basic question: how long can it last?

The years after the Second World War were perhaps the first time in history when the physical limits of our planet became clearly recognizable for everybody. It was past the time when maps showed white areas with the writing, "*hic sunt leones*"; "there are lions here." Already in 1953, Edmund Hillary and Tenzing Norgay had climbed to the top of Mount Everest, the highest mountain on Earth; one of the last spots on the planet's surface that had remained untouched by humans. In the 1960s, photos of the earth from space, brought back by the astronauts of the Apollo space-ships, showed images of our planet as a blue-green ball. It was finite, it was limited; everybody could see that. "No man is an island", John Donne had said. It was now easy to understand that we were all living on a small blue island floating in the blackness of space.

The idea that all living beings were part of the same entity goes back to the 1920s, when Vladimir Vernadsky had coined the term "biosphere" (Weart 2003). The concept became widely known in the 1960s, when James Lovelock developed the concept of "Gaia" (Lovelock 1965), borrowing the name of the ancient earth divinity to describe the planetary ecosystem. Gaia embodied the concept that all the creatures of the Earth are linked in a complex system of feedbacks that maintain the temperature of the planet at levels tolerable for life.

In time, Gaia turned out to be far different from the benevolent entity that Lovelock had imagined (Ward 2009). But the concept of Gaia remains a valid metaphor even today for the description of the planetary ecosystem (Karnani and Annila 2009). Then, if life is a single, giant organism, it follows that this organism is tied to the limits of the whole planet. And, if the bounty of the planet is limited, how could we continue forever the rapid growth experienced in the 1950s and 1960s?

In 1968, Garrett Hardin provided more food for thought on this problem with his paper "The Tragedy of the Commons" (Hardin 1968). Hardin described a case in which land is a common good, that is, it is free for all shepherds as pasture. Each shepherd understands that too many animals damage the land. Yet, each shepherd tends to add more animals to his own herd because his individual advantage is larger than the dam-age caused to him by the collective loss. The result is disaster: with too many animal grazing, the fertile soil is destroyed by erosion and everyone suffers as a consequence. Still, even though the situation is evident to everyone, it remains convenient to each individual shepherd to overexploit even the last patches of pasture remaining.

Hardin's model was schematic and, surely, the real world is much more complex than the hypothetical pastures he had described. Nevertheless, with his analysis, Hardin questioned at its core the concept of the "invisible hand" that Adam Smith had developed almost two centuries before. Whereas neoclassical economists saw the pursuit of self interest as leading to the maximum common good, Hardin saw it as leading to the destruction of the resource being exploited.

While Hardin's qualitative model explained the reasons for the overexploitation of natural resources, in the 1950s, the American geologist Marion King Hubbert developed an empirical model that described how fast resources were exploited. Hubbert proposed that the production curve for crude oil in any large productive region should be "bell shaped." Applying his model to crude oil in the 48 US lower states, he predicted in 1956 that maximum production should have been attained around 1970 and that it should have been declining afterwards. That was what happened, with the peak reached in 1971 (Campbell and Laherrere 1998).

Today, the "bell shaped" production curve has been observed for many minerals other than crude oil (Bardi and Pagani 2007) and even for slowly renewable, nonmineral resources (for instance for whales (Bardi 2007b)). The bell shaped curve is known also as the "Hubbert curve," while the maximum is often called the "Hubbert peak" (Deffeyes 2001, 2005). The worldwide peak of oil production is often called "peak oil" (Campbell and Laherrere 1998).

Crude oil in the US was not the first case of a major mineral resource having reached its production peak. Earlier on, around 1920, British coal production had also peaked and had started its decline (Bardi 2007a). These and other historical production peaks passed largely unreported and, whenever discussed, were normally attributed to market factors. Nevertheless, these data showed that resources, even important ones, were finite and that production could not be kept increasing forever.

Another worry of those times was the growth of human population; a crucial factor of Hardin's "commons" model (Hardin 1968). Up to then, population growth had been always seen as a good thing. More people meant more soldiers for the army, more taxpayers for the state, more peasants for landlords, more workers for factory owners; in short, more wealth for everyone. But, in the 1950s and 1960s, population growth had taken an ominous trend. At the beginning of nineteenth century, the human population had reached one billion. In 1927, it had touched two billions. The three billion mark had been reached in 1960s. Growth seemed to follow an exponential curve and it was a nightmare to extrapolate the trend to future decades. Worries about overpopulation were diffused in a popular book by Paul Ehrlich "The Population Bomb" (1968) where the author predicted widespread famines all over the world.

Finally, the world "pollution," so familiar to us, is recorded to have been used with the meaning of "environmental contaminant" for the first time in the 1950s. Initially, the term was referred mainly to the radioactive waste generated by nuclear explosions but, with time, there started to appear other forms of pollution. "Smog," the fumes resulting from the combustion of coal became a deadly threat. The 1952 London smog episode was especially deadly, with about 4,000 victims (Stegerman and Solow 2002). This event generated legislation against pollution: the Clean Air Act, enacted in Britain in 1956 and the "Air Pollution Control Act" enacted in the United States in the same year. Later on, another, more comprehensive piece of federal legislation in the US, the "Clean Air Act," was enacted in 1963.

Pollution took many forms, often unexpected. One was the effect of pesticides. In 1962, the American biologist Rachel Carson published her book "Silent Spring" (Carson 1962) where she criticized the indiscriminate use of pesticides; that she termed "biocides." Carson's claim was nothing less than revolutionary. Up to that time, an obvious truth had been that living creatures could be divided into "useful" and "noxious" and that it was a good thing to exterminate the latter kind. The concept of "ecosystem," where many species interact with each other, actually need each other, was relatively new in a world dominated by concepts such as "better living through chemistry." Carson was probably the first author to describe these issues in a popular book and the resonance was enormous. She is correctly credited as having started what we call today the "environmental movement."

These years also saw the first hints that human activity was creating an even more worrisome threat to humans: global warming. In 1957, Robert Revelle and Hans Suess (Revelle and Suess 1957) took up earlier work on the effect of greenhouse gases and published a study that evidenced how the observed growth of carbon dioxide resulting from human activities would result in a rise in the planetary temperatures. The evidence of global warming was not to become completely clear before the 1970s but, already in the 1960s, it was another worrisome indication of the effect of humans on ecosystems.

In the 1950s and 1960s, worries about resource depletion, overpopulation, and pollution were not at the top of the daily news, but were slowly making their way inside the consciousness of many people concerned about the fate of humankind. Something was to be done, clearly, but it was not enough to say that, someday, we would have to come to terms with our environment. The point was *when*. Did we have millennia, centuries, or perhaps just decades? And not just that; *how* would the clash of humankind against planetary limits take place and what symptoms could alert us and help us preventing the worst consequences? There was a tremendous need to understand what the future had in store for humankind.

Quantifying this kind of global problems had never been attempted before, but the progress of technology was starting to provide the right kind of tools for such a purpose. In the 1950s and 1960s, digital computers were rapidly developing and in several areas of science they started to be utilized as instruments for forecasting the future.

In the late 1940s, a young professor at the Massachusetts Institute of Technology in Boston, Jay Wright Forrester, had started using a new generation of digital computers to simulate the interactions of different elements of machinery and electronic systems. In time, Forrester moved his interest to management and to the modeling of economic systems. He dubbed the new field "industrial dynamics" (Forrester 1958, 1961) but, later on, the term "system dynamics" became commonplace. The method could be used to simulate business situations (Sterman 2002) and socioeconomic systems such as entire cities (Forrester 1969; Madden 1979; Burdekin 1979). From there, another step forward was clear for Forrester: simulate the world's entire economic system. He developed his first world models in the mid 1960s, when his computers had become powerful enough to carry out simulations involving the main elements of the world's economy and of the ecosystem.

While Forrester was engaged in these studies, others were examining the same problems. In 1968, Aurelio Peccei (1908–1985) and others formed the "Club of Rome," a group of intellectuals from industry and from academia who had gathered together to study and discuss global economic and political issues. At the beginning, the focus of the Club was not specifically on planetary limits. The document titled "The Predicament of Mankind" (Club of Rome 1970) shows that the main concern of the Club at that time was finding practical ways to improve the conditions of life of human beings; in particular by reducing inequality in the distribution of wealth. But the members of the Club rapidly came to the conclusion that they had to find ways to quantify the limits of the world's resources if they wanted to be able to act on their concerns. In 1968, Aurelio Peccei met Jay Forrester at a conference on urban problems in Italy, on the shores of Lake Como (Forrester 1992). That meeting was the spark that ignited the series of events that would lead to "The Limits to Growth" study.

Peccei was impressed by Forrester's ideas and he invited him to participate at a meeting that the Club of Rome organized in Bern in 1970. At the meeting, Forrester persuaded the members of the executive committee of the Club to travel to MIT, in Boston, to discuss the possibility of using system dynamics for their purposes. It is said that Forrester jotted down the basic elements of his world model while flying back to Boston.

In Boston, the members of the Club of Rome discussed with Forrester and also with one of his coworkers, Dennis Meadows, then 28 years old. Apparently, Meadows was the only one in Forrester's group who was not already burdened with other projects. So, he wrote a memo of several pages in which he outlined how a world modeling project based on system dynamics could be run and the members of the Club decided to try to use it. Later on, Meadows, following Peccei's suggestion, applied to the Volkswagen foundation to provide a grant for the study (Meadows, private communication, 2010). The proposal was accepted and Meadows assembled a team of 16 researchers to work on the project. The team included Dennis Meadows' wife, Donella Meadows, Jorgen Randers, William Behrens, and others.

In 1971, Forrester published the results of his work on world modeling in a book with the title "World Dynamics" (Forrester 1971). The work of the team headed by Dennis Meadows was published in 1972 as a book with the title of "The Limits to Growth" (LTG) (Meadows et al. 1972).

Forrester and the LTG team had worked independently of each other, but had arrived at the same shocking conclusion that can be summarized as: *The world's economy tends to stop its growth and collapse as the result of a combination of reduced resource availability, overpopulation, and pollution.* This conclusion was a "robust" feature of the simulations; that is, it changed little when varying the initial assumptions.

Neither Forrester's calculations nor the LTG ones were meant to determine when exactly the collapse was to start but, using the best available data, both studies indicated that the start of the economic decline could be expected within the first decades of the twenty-first century, that is about 30 or 40 years into the future. In the LTG study, this scenario was referred to as the "Base case" or the "Standard Run" (Fig. 2.1).

Both Forrester and the LTG team performed their simulations for a variety of possible assumptions, including radical technological innovations or that population could be stabilized by policy actions at the global level. In most cases, even for very optimistic assumptions about resource availability and technological progress, collapse could not be avoided but, at best, only delayed. Only a carefully chosen set of world policies designed to stop population growth and stabilize material consumption could avoid collapse and lead the world's economy to a steady state while maintaining an average level of material consumption not different than it was in the 1970s.

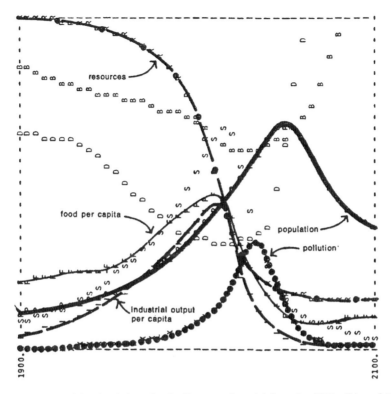

Fig. 2.1 Results of the simulations for the "base case" model from the 1972 edition of "The Limits to Growth." The base case model is the one which assumes as input parameters the values closest to the available data. Image credit: Dennis Meadows

The two books describing these concepts, "World Dynamics" and LTG, both had a considerable success. Forrester's book sold about 100,000 copies, a remarkable result for a book that was conceived as a technical text and that was full of equations and diagrams. But the real impact came with the LTG study, which was aimed from the beginning at the general public.

The total number of copies of LTG sold is not known with certainty. However, the authors of the study (Meadows, private communication, 2010) estimate that the total sales were surely over one million. The book was also translated into more than 30 languages. Evidently, the book gave voice to something that was deeply felt all over the world: that the limits to the planetary resources could not be ignored for long.

In spite of the success and the interest raised, the work of Forrester and of the LTG team also encountered strong criticism. Economists, for instance, seemed to be unified in rejecting the methods and the findings of these studies as not compatible with economics as it was understood at that time. But where the debate really raged was in political terms. Here, the very success of the LTG book generated a strong reaction.

The reviews by Giorgio Nebbia (1997) and by Mauricio Schoijet (1999) tell us how different political attitudes shaped the worldwide reaction to LTG. In the Soviet Union, the reaction was that the book might have well described the collapse that was in store for Capitalism, but that it had nothing to do with Communist societies which would avoid collapse by means of their planned economies. In many poor countries, local intellectuals often saw LTG as an attempt to perpetuate the dominance of the rich West, a fraud to impose population reductions on the poor or even the harbinger of a return to colonialism. In the Western world, different political orientations often dictated the reaction to LTG and, in many cases, left and right were united in the rejection. The left often saw LTG as an attempt to justify the subordinate position of the working class while the right saw it as a threat to their vision of free markets and economic growth. Positive political reactions to LTG came most often from moderate–liberal positions, which not only saw the threats described in the LTG scenarios as real but also as opportunities to reduce inequality and create a freer world (see, e.g., Peccei and Ikeda 1984).

If, at the beginning, the debate on LTG had seemed to be balanced, gradually the general attitude on the study became more negative. It tilted decisively against the study when, in 1989, Ronald Bailey published a paper in "Forbes" where he accused the authors of having predicted that the world's economy should have already run out of some vital mineral commodities whereas that had not, obviously, occurred.

Bailey's statement was only the result of a flawed reading of the data in a single table of the 1972 edition of LTG. In reality, none of the several scenarios presented in the book showed that the world would be running out of any important commodity before the end of the twentieth century and not even of the twenty-first. However, the concept of the "mistakes of the Club of Rome" caught on. With the 1990s, it became commonplace to state that LTG had been a mistake if not a joke designed to tease the public, or even an attempt to force humankind into a planet-wide dictatorship, as it had been claimed in some earlier appraisals (Golub and Townsend 1977; Larouche 1983). By the end of the twentieth century, the victory of the critics of LTG seemed to be complete. But the debate was far from being settled.

The interest in LTG and, in general, in world modeling, may be returning today as the collapse seen in all the scenarios of LTG may start to manifest itself in terms of crashing markets and economic crisis. With the new century, we are seeing a growing realization that we need new ideas and approaches to address severe economic and environmental problems from resource depletion to global climate change.

Today, we can look back at almost 40 years of the story of "The Limits to Growth" and review the message that we received in 1972. What was the future at that time is now the past and we can recognize the validity of the scenarios presented in the book (Turner 2008; Hall and Day 2009) with the world's economic system having closely followed the LTG "base case" scenario up to now. So, we can say that LTG never was "wrong" in the sense that critics intended. They had badly misunderstood, forgotten, or ignored that the time scale of the base case scenario was such that collapse was not to be expected until a time that was roughly estimated as within the first two decades of the twenty-first century.

There is also a deeper factor in the misunderstanding the LTG message. It was a warning, not a prediction and, as such, it could not be "wrong." It does not matter whether collapse occurs or not on the specific date that corresponds to a specific scenario of those presented in the book. What matters is that, by ignoring the study for the past four decades, we probably lost the ability to do something to avoid, or at least mitigate, the problems ahead.

The story of "The Limits to Growth" is a continuing one. We can still learn a great deal from the 1972 study and from its more recent versions (Meadows et al. 1992, 2004). So, it is not too late to put into practice some of the methods and recommendations that can be derived from the study.

Chapter 3
Of Models and Simulations

Science is mostly about making predictions. We make measurements and we build models (that we also call "theories" or "laws") all in order to make sense of what happens around us and to face the future without being unprepared for it. In terms of its ultimate purpose, science has not changed very much from the time of ancient oracles although, of course, today we are proud of defining some of our sciences as "exact."

With "The Limits to Growth," a problem that appeared very early in the debate was that of understanding exactly what were the purposes of the study. The insistence of the authors in saying that their results were "not predictions" generated much confusion. If these are not predictions, some said, what is the study good for? Others took the results of the study as forecasts, if not as oracles, and tended to select those scenarios which they liked (in order to praise the study) or which they disliked (in order to criticize it).

So, if we want to understand the relevance of the LTG study, we need to understand first what kind of systems it was meant to model, with what aims, and with what kind of expectations. Models "can be made and judged only with respect to a clear purpose" (Meadows et al. 1982). This is what the present chapter is dedicated to.

Modeling is an ancient art and the first models of complex systems may go back as far as the Chinese strategist Sun Tzu, who wrote in his "The Art of War," (ca. 400 BCE Project Gutemberg 2010)

> The general who wins a battle makes many calculations in his temple ere the battle is fought. The general who loses a battle makes but few calculations beforehand. Thus do many calculations lead to victory, and few calculations to defeat: how much more no calculation at all!

With the term "calculations," we may assume that Sun Tzu was meaning that military planners should make quantitative predictions of the future. He may have been referring to something similar to what we call today "military simulations" or "wargames."

It is not casual that the first examples of modeling in history appear in the military field. War is a complex and difficult activity, fraught with uncertainties and danger; and, its final outcome strongly depends on decisions made while the system evolves. Modeling such a system is extremely difficult: if it were possible to predict exactly the outcome of wars, then it would be pointless to engage in one!

Nevertheless, some degree of prediction is always possible, even for wars and battles. Then, of course, some degree of prediction is better than no prediction at all. That is true for military simulations, but it is of general validity in the whole wide field of modeling. According to the definition given by Hall and Day (1977) "Models are a formalization of our assumptions about a system." Our assumptions, obviously, can be wrong and, in this case, the purpose of the model is to show us our mistakes. In any case, models allow us to make explicit how we think things work and to apply the scientific method to complex issues.

In general, models can be purely qualitative, based solely on intuition and experience. Models can also be quantitative, based on mathematical expressions that define how the various elements of a system are related to each other. For quantitative models, "running a simulation" means that the model's equations are solved stepwise as a function of time; usually using a computer. This procedure provides a description of how the parameters of the system vary with time. When a simulation captures just one of many possible different outcomes, then the term often used is "scenario."

The systems that can be modeled have been classified in many different ways (Ackoff 1971; Lewis 2002) depending on their structure, purpose, organization, etc. We can broadly divide systems into two categories: *linear* and *nonlinear*. It would be out of scope here to go into the details of the exact definitions of these two terms. However, we can say that linear systems are the domain of traditional physics and engineering, whereas nonlinear systems are those studied mostly by biology, sociology, and economics.

Linear systems are characterized by well-defined cause and effect relationships and their behavior is predictable and repeatable in time. A good example is a mechanical clock, whose inner gears interact with each other in a predictable manner. We have obtained remarkable success in understanding and predicting the behavior of these systems although, obviously, there are limits to how exactly the behavior of linear systems can be predicted. For instance, the trajectory of a projectile is influenced by effects, such as wind speed, that cannot be exactly determined before the projectile is shot. Nevertheless, a flying body in a gravitational field is a case of a linear system so that, on the whole, prediction is possible with a degree of approximation that can be improved by using more sophisticated calculation methods and better estimates of the parameters involved.

But there exists another category of systems, that of *nonlinear* systems. These systems usually involve several bodies or elements that interact with each other. They show typical characteristics; in particular, their evolution in time is often very difficult – or impossible – to predict over an extended time range. They also do not show a well-defined cause-effect relationship.

Nonlinearity is a fundamental element of the category of systems defined as "complex." In this context, the term "complex" does not mean simply complicated, in the sense that the system is formed of many elements or subsystems. It means that the elements of the system are linked to each other in ways that usually involve "feedback" relations. As a general definition, feedback means that the behavior of a system is determined by the past history of that system. But the concept of feedback as it is intended in nonlinear systems implies that the elements of the system react to a perturbation amplifying it or dampening it depending on whether the feedback is positive or negative. "Negative" and "positive" in this sense have no normative (good or bad) intrinsic meaning. This point is often difficult to understand for many people and, hence, the two terms are sometimes referred to as "reinforcing feedback" and "balancing" or "stabilizing" feedback. Perturbations coming from outside the system are normally called "forcings." A forcing, typically, generates a cascade of feedbacks within the system that may lead to a strong amplification, or dampening, of the perturbation.

Complex, interactive, and nonlinear systems often show a set of typical characteristics. These properties have been described by Sterman (2010) as:

- Dynamic
- Tightly coupled
- Governed by feedback
- Nonlinear
- Self-organizing
- Adaptive
- Evolving

This list gives us some idea of the fascination involved with the study of these systems and also of the difficulty in making prediction of how exactly they will behave.

A simple system that shows feedback effects is the governor of a steam engine that was described first in mathematical terms by James Clerk Maxwell (1868). The governor is a typical example of negative (stabilizing) feedback. When an external forcing causes an increase in the speed of the engine; the governor reduces the flux of steam to the engine, slowing it down. The opposite occurs for a decrease of the engine speed. These stabilizing feedbacks maintain the engine at a constant speed. The governor/engine system shows a property called "homeostasis," the tendency of maintaining its parameters constant or oscillating around constant value. This property is typical of complex systems where stabilizing feedbacks are at play (Karnani and Annila 2008).

A more complex example of a feedback dominated system is the earth's atmosphere, which includes both stabilizing and reinforcing feedbacks. In normal conditions, the atmospheric temperature is in a homeostatic condition and changes very little over time, apart from seasonal cycles. But emissions of CO_2 generated from human activity are a forcing that unbalances the outgoing and incoming flux of solar radiation by affecting the "greenhouse effect" associated with some of the components of the atmosphere. The overall result of an increase in CO_2 concentration must be a warming effect, but the climate system reacts to this forcing with a series of positive and negative (reinforcing and stabilizing) feedbacks that may amplify or dampen the CO_2 forcing. An example of stabilizing feedback is that of plants, which will increase their growth if there is more CO_2 in the atmosphere and thereby tend to stabilize its concentration. An example of reinforcing feedbacks, instead, is the effect of the changing extension of glaciers which affects the planetary albedo. With increasing temperature, glaciers tend to shrink in their area; that reduces the planetary albedo and causes a further increase of temperature. The earth's climate system is a tangle of feedbacks which will eventually unravel as the result of the forcing imposed by CO_2 emissions. The result will be what Robert Luis Stevenson called "a banquet of consequences" (Sterman 2010) which might not be pleasant at all for human beings (Hansen 2007). The fact that we are continuing to emit greenhouse gases despite the dangers involved shows how difficult it is to predict the behavior of nonlinear systems and how most people have troubles in understanding the consequences of actions performed with such systems.

Some nonlinear systems show a further degree of complication: becoming "chaotic" (that is, never repeating the same trajectory in time) or showing what has been called "self-organized criticality" (SOC) (Bak 1996). SOC systems can be described by means of "power laws" that link the intensity of a certain event to its frequency in time.

A system that is chaotic or in a SOC state may be "deterministic," in the sense that it is generated by apparently simple equations, but exactly predicting its behavior is very difficult. Even a simple logistic equation, $x(t+1) = kx(t)(1 - x(t))$, becomes chaotic and unpredictable for some values of k. This equation is known in ecology as the "May equation."

However, the fact that a system is dominated by feedback effects, or even that it is chaotic or continuously evolving, does not imply that it is completely random. On the contrary, these systems often follow well-defined patterns. Think of a biological system: clearly it is impossible to devise an equation based on Newton's laws that would describe the motion of a live cat. But a cat's behavior is not completely unpredictable: just rattle the box of the dry food and see what happens.

For these systems, one often speaks in terms of "attractors," sets of parameters that the system tends to attain. In some cases, the system will actually attain these parameters and there it will settle.[1] In others it will "orbit" around one or more attractors, without ever reaching one of them. In the case of a chaotic system, the orbits will never repeat previous trajectories and, in this case the attractors are called "strange attractors." A classic example of strange attractor is that related to the so called "butterfly effect." The term originates with a paper by the meteorologist Edward Lorenz (1963) that has been picked up in popular culture – as in the 1993 movie "Jurassic Park." Whether such effects are in fact operational is quite another question, but the butterfly effect is a useful concept in order to illustrate the fact that nonlinear and complex systems do not show a well-defined cause-effect relationship.

---

[1] The box of the dry food could be defined as an "attractor" for the cat, although not with the same formal definition of attractors in chaos theory.

The butterfly does not "cause" a hurricane, it is just a trigger that unchains the series of events leading to it.

So, predicting the behavior of complex, nonlinear systems is much more difficult than for the case of linear ones. Nevertheless, some degree of prediction is possible, provided that we do not pretend to have both exact and long-term predictions. A short-term prediction of a nonlinear system falls in the category that includes weather forecasts. In this case, no matter how sophisticated the models used are and how extensive is the database considered, the model and the real system will rapidly diverge with time. Weather forecasts, as we all know, are valid for a few days at most. Nevertheless, they are still useful within that short time span.

Long range predictions, instead, fall in the category of climate modeling. This kind of prediction has to do with determining the attractors of the system. Here, we do not need mathematical models to know that any given day in August will normally be hotter than any day in January. But, of course, we cannot predict in January exactly how hot a specific day of the following August will be. Detailed climate models are used for general predictions of long range trends (IPCC 2007) and have a long history (Richardson 1922; Ashford 1985). Despite their sophistication, however, these models are unable to tell you whether tomorrow it will be raining or not. So, it is a general rule that applies to the predictive ability of these systems that we must choose between range and precision.

At this point, it is possible to understand what kind of system LTG was modeling. The world's economic system is complex, nonlinear, and feedback-dominated. So, it is impossible to make detailed predictions over decades or even just years. Nevertheless, it is still possible to understand the behavior of the system over these time scales. The world system modeled in the LTG study does not show chaotic behavior and is not cyclical (although it may be, over a longer time range than the one examined). Therefore, we cannot define "attractors" for this system. However, it is possible to model the system's behavior for a range of sets of initial assumptions and determine what features can be defined as "robust"; that is do not depend, or weakly depend, on the input parameters or on external forcings. As mentioned earlier, for the cases of LTG and of "World Dynamics," it could be determined that collapse of the world's economy is a robust feature of the model.

These world models are therefore analogous in scope and approach to recent studies on climate (IPCC 2007), where the results of the simulations are reported in terms of various possibilities that depend on the initial assumptions. None of the climate scenarios discussed in the IPCC reports is a "prediction," but – taken together – they tell us that warming is a robust feature of the system being modeled. However, just as for the case of LTG models, climate models have been much misunderstood and maligned, with the inherent uncertainty which is part of scenario building being taken as a weakness of the whole concept of "global warming."

From these considerations, we can now list what models and simulations can be used for, with a special view on nonlinear, complex systems.

1. *Prediction*. Models, even of complex, nonlinear systems, have a certain predictive value. Of course, all predictions should be taken with caution. Most simulations depend on a large number of input parameters and each parameter has a specific

uncertainty associated with its value. Uncertainties exist also regarding the relations that describe the model. Furthermore, the results of the simulation will normally be affected by decisions taken while the system evolves (it is well known that no battle plan survives contact with the enemy). Because of these uncertainties, predictions obtained using simulations should always be taken as approximate. Nevertheless, even within these limits, predictions are not useless. For instance, in most cases, a military simulation will generate the victory of the stronger side; thus confirming Napoleon's statement that God fights on the side with the best artillery.

2. *Exploration*. This is the concept of running a "fan" of scenarios that will cover most of the reasonable possibilities for the future to manifest itself. This is a typical military approach, consisting in laying out the range of possible strategic outcomes which will unfold depending on the choices made by the enemy and by one's side. Of course, this kind of approach is not limited to military simulations but can be applied to all sorts of systems, in business, economics, urban planning, and many others. Exploring the possible options is the first step in controlling a system.

3. *Training*. This is a typical use of simulations, military, and otherwise. Running the simulation allows the people responsible for the control of the system to gain familiarity with their job and on how their actions affect the behavior of the system.

4. *Preparation and planning*. This use of the models consists in making actual use of the results of simulations to prepare for the future. In a strategic simulation, it may consist on deploying military units in such a way as to optimize the response to different possible actions by the enemy. The same strategy can be used in business or for tackling pollution problems such as climate change. You can never know exactly what the future has in store, but you can be prepared for it.

That is what models and simulations are, but it is worth also listing what they are *not*. Simulations are not, for instance "just a game," notwithstanding the fact that military simulations are called "wargames." It may be that the term "game" has a negative ring, but wargames have been shown over and over to be a fundamental military tool (Caffrey 2000).

Furthermore, models and simulations have nothing to do with what is called "sympathetic magic," the concept that what you do to a representation of something will affect the thing being represented. For instance, within the paradigm of sympathetic magic, sticking pins into a doll that looks like someone (the "model") is supposed to make that someone feel pain or fall sick. Needless to say, this is not the way computer-based models and simulations are supposed to work! Nevertheless many people seem to believe that the result of a simulation affects the behavior of the real world.

But simulations are mathematical models; equations that generate numbers and graphs shown on a computer screen. Simulations describe situations that we may recognize as possible outcomes of a real world system and that is all. Simulations may be more or less useful, but they cannot be, strictly speaking, "right" or "wrong."

Yet, people's emotional attitude will cause them to see simulations exactly in these terms and they will become angry or happy depending on the results of a simulation. That is one of the problems that LTG encountered: many people were upset because it was understood that the study was actually predicting a catastrophe or, worse, it was a piece of sympathetic magic that would *cause* a catastrophe (Marxsen 2003).

It takes some effort to free oneself from this kind of biased perceptions. Once that is done, it is possible to benefit from models and simulations as a useful, actually indispensable, tool for being prepared for the future.

Chapter 4
System Dynamics Modeling

System dynamics had an adventurous beginning. Jay Wright Forrester, who developed it in the 1950s, had started his career by studying servo systems for automated antiaircraft guns. This interest led him to travel to the Pacific Theater, during the Second World War, to repair an automatic radar system installed aboard the aircraft carrier Lexington, which was torpedoed by the Japanese while he was on board.

After the war, Forrester continued to work on servo systems at the Massachusetts Institute of Technology in Boston. There, he led a team that developed a magnetic memory that became the standard for computers in the 1950s (Forrester 1951). Forrester soon applied the new computers to the study of nonlinear systems and, in 1956, he moved to the Sloan School of Management at MIT where he started working on applications of the methods to social and economic systems. In time, he developed the set of tools and concepts that are known today as "system dynamics" (Forrester 1992).

System dynamics is a way of modeling any kind of systems, but is especially thought to be used for nonlinear, complex systems. The term "dynamical" may simply mean "time dependent," but it was chosen by Forrester (1992) to stress that these models did not simply aim at creating a snapshot of the system being modeled, but to describe its evolution in time.

System dynamics models take into account a number of fundamental elements; mainly *stocks, flows and feedback*. A stock is an *amount* of something; for instance it can be the number of individuals in a population. The *flow*, instead, is the variation of a stock with time; for instance, in a biological system it may be the growth (or decrease) of a population. Flows and stocks are related to each other: the size of a stock will always vary depending on the incoming and outgoing flows, while the intensity of flows will often depend on the size of a stock it is connected to. When the intensity of a flow into a stock depends on the size of the stock, we have an operational definition of "feedback." In this case, stock and flow influence each other, reinforcing or damping the growth or the decline of the stock. Positive feedback is defined as enhancing the process while negative feedback as restricting the process. The equivalent terms "reinforcing feedback" and "stabilizing feedback" are also used.

U. Bardi, *The Limits to Growth Revisited*, SpringerBriefs in Energy: Energy Analysis, DOI 10.1007/978-1-4419-9416-5_4, © Ugo Bardi 2011

System dynamics is a way of quantifying these factors in order to describe the behavior of complex systems. As mentioned before, a complex system does not have to be very complicated, so we may start with a simple example in order to describe how system dynamics works in practice.

A simple model that includes both negative and positive feedbacks is the one that today takes the name of "Lotka-Volterra" (LV) (Lotka 1925; Volterra 1926). This model is also referred to as the "predator–prey" model and, sometimes, as the "wolves and rabbits" or "foxes and rabbits" model. The LV model is an extreme simplification of an actual biological system and it has been shown to be unable to describe actual biological populations (Hall 1988). Nevertheless, we can consider it as a mental tool (a "mind sized" model; (Papert 1980)) useful to understand the main features of more complex dynamic models.

Using the terminology common in system dynamics, we can define the two stocks in the LV model: the population of prey and that of predators. Let's call R the stock of the resource (the prey) and C the stock of predators (later on, we'll call this entity "capital" when applying the model to economic systems). We call R' the variation (flow) of the stock of prey and C^c the variation (flow) of the predators. With this notation, the LV model can be written as two coupled differential equations, where the "ks" are positive coefficients to be adapted to the specific system being modeled.

1. $R' = -k_1 CR + k_2 R$
2. $C' = k_3 CR - k_4 C$

Note how these equations describe feedback effects within the system. Every term in the equations shows a feedback relationship; that is, the flow of each stock depends, among other factors, also on the size of the stock. For instance, the growth of the rabbit population depends on the positive (reinforcing) feedback associated with the $k_2 R$ term: the growth rate is proportional to the number of rabbits – more rabbits mean more growth. If rabbits were not culled by foxes, their number would go up exponentially forever (at least in the model). Note also the negative (stabilizing) feedback that links the number of rabbits and the number of foxes. The term $-k_1 CR$ indicates that the more foxes ("C") there are, the faster the number of rabbits tends to shrink.

These considerations tell us something about the relations among the two elements of the system. But, just by looking at the equations, we cannot easily grasp how the system will behave when all its elements interact together. To do that, we must solve the equations, that is run a simulation in which the various stocks and flows of the system are determined stepwise as a function of time.

The coupled system of equations of the LV system does not have an analytical solution, but it can be solved iteratively. The result is a pattern of oscillations in the number of prey and predators (foxes and rabbits) as seen in Fig. 4.1.

In the simulation, the interplay of feedbacks makes predators and prey oscillate in phase with each other. Neither the number of foxes nor those of rabbits tend to go to zero or to shoot to infinity. When there are too many rabbits, foxes grow to cull their number. When there are too many foxes, they die of starvation.

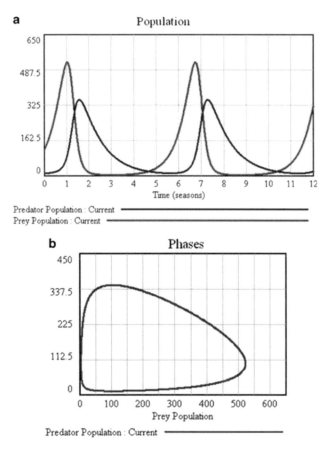

Fig. 4.1 Results of the simulation of a simple model; the "Lotka-Volterra" (LV) model. The model describes a simplified biological system where two species (predator and prey) interact. The result is oscillations around a constant value as the results of the feedback relationships within the system (Calculations made using the Vensim (TM) software package)

Note how (Fig. 4.1b) the number of rabbits varies as a function of the number of foxes: the plot is a closed loop that repeats forever. We could say that foxes and rabbits "orbit" around a point that we may call the attractor of the system. This attractor is a couple of values of the two populations which – if attained – stabilize the system. That will happen in the LV model if the equations are modified by adding a "friction" term that, eventually, will cause the number of individuals of both species to stabilize to the value of the attractor. But, in the simple case described here, the orbiting and the oscillations will go on forever.

The LV model is, of course, a very simplified way to describe how an ecosystem might work, but it does catch something of the interplay of different biological species in the real world. Note, among other things, that it shows homeostasis; that is it tends to maintain its parameters oscillating around fixed values. This property

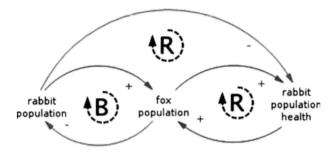

Fig. 4.2 Causal loop diagram of the LV system. In the figure, "R" indicates a "reinforcing" (positive) feedback and "B" indicates a "balancing" (negative) feedback. Image credit: SystemsWiki – www. systemswiki.org/

derives from sound thermodynamic principles that operate on systems subjected to external fluxes of energy (Karnani et al. 2009).

The LV model may be seen as a starting point to understand models of socio-economic or biological systems. Typically, these systems require a large number of differential equations and more parameters. Using computers, it is easy to use iterative methods to solve even large number of coupled differential equations. The problem is that for the unaided human mind it soon becomes very difficult to manage a multielement model simply looking at the relative equations. For this reason, graphical representations of system dynamics models are normally used.

The modern graphic representation of dynamic systems has been somewhat simplified compared to the early versions; although some elements remain the same. Stocks are drawn as squares; flows are shown as arrows. Positive and negative feedbacks are shown with + or – signs, sometimes enclosed into parentheses as "(+)" and "(−)." Sometimes, the letter "R" is used to indicate a reinforcing (positive) feedback and the letter "B" for a balancing (or negative) feedback. A more complex set of symbols describing this kind of systems has been designed by Howard Thomas Odum (1994). In this case, there are specific symbols for concepts such as "production," "switch," "energy loss," and others. This set of symbols and conventions is sometimes referred to as "Energese," a form of language (Brown 2004). However, here we will consider only the standard system dynamics conventions.

The LV model can be described using system dynamics symbols in what is normally called a "causal loop diagram," that is a diagram that illustrates the basic interactive relations of the various elements of the model (Fig. 4.2). Shows a simplified version of the LV model that emphasizes the feedback relationships.

Causal loop diagrams can be used for a first, rapid understanding of the structure of a model. These diagrams are qualitative schemes that cannot be used to solve the equations of the model. It is possible, however to use computer-based graphical interfaces which not only visually describe the model, but are also linked to the relative equations and permit users to solve them in an interactive manner (e.g. Wilensky 1999). One such interface for the LV model is shown in (Fig. 4.3).

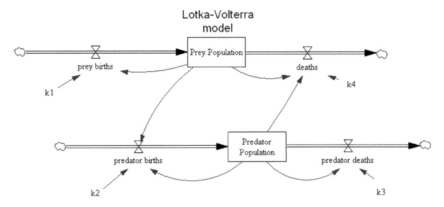

Fig. 4.3 The LV model shown as a graphical interface according to the graphic conventions of system dynamics as interpreted in the "Vensim™" software package

In this kind of graphical interface, stocks are indicated by squares, as in causal loop diagrams, but flows are defined by double edged arrows which include an element called the "valve" that regulates the flow. Single edge arrows indicate the feedback relationships, just as in casual loop diagrams.

In the case of the LV model, the stock of prey (the square) is connected to the flow (the "valve") through a double edged arrow. The single edged arrow goes back from the stock to the flow and indicates the feedback. Similar relations exist for the predator stock, which appears as another box.

Software packages that provide this kind of interface allow users to solve the model without the need to see the actual equations. However, it is also possible to toggle back and forth from the graphic image to the equations. There are several such commercial packages on the market today. In many cases, the software offers users sophisticated interactive methods to vary the parameters of the systems while observing how the system behavior varies as a consequence.

Using modern software packages it is possible to deal with complex dynamic models that can be used to describe a variety of business, social, economic, and other types of systems (and to make mistakes if you are not very careful). In the figure, an example of a multielement system is shown, the nitrogen cycle of the earth's ecosystem (Fig. 4.4).

The results of simulations are often expressed in the form of graphs or tables, but it may also be communicated to users in terms of dynamic maps and animations. A special kind of user interface is the so called "flight simulator" where users can practice how different choices in the input parameters can affect the results of the simulation, learning how their own decisions affect the short- and long-term behavior of the system. Such an interface is shown in Fig. 4.4b for the nitrogen cycle (from www.systemswiki.org). A similar method has been developed to study the effects of greenhouse gas emissions on the earth's atmosphere and the effects of policy options to reduce emissions (Sterman and Sweeney 2007).

a

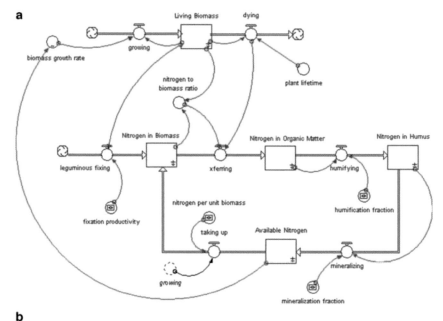

b

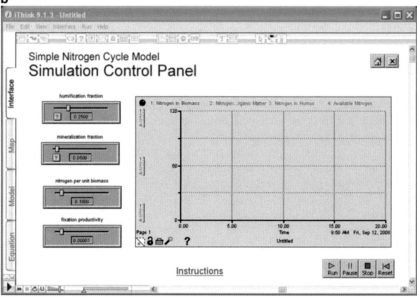

Fig. 4.4 An example of a multielement system dynamics model, the earth's nitrogen cycle simulated using the "Stella" software. (**a**) The model; (**b**) the "flight simulator" panel, for users to understand the effect of varying the model's parameters from www.systemswiki.org

System Dynamics in the form of linked differential equations managed by graphical interfaces remains today probably the most common method of studying complex, non-linear systems. It is not, however, the only possible method. An alternative is "agent based" (AB) modeling, also called ABS, "agent based simulation" or MAS, "multiple agent simulation" (see. e.g., (Shoham and Leyton Brown 2008)).

AB simulations are a way of modeling the system under study at the level of single agents, which might be single human beings in the case of a socioeconomic system. AB models may have a higher degree of detail than system dynamics models and usually include such features as random noise. Hence, AB models may offer more possibilities than system dynamics ones but also need more parameters which may not be exactly known. In several cases, the same systems can be described with equivalent results by AB and by system dynamics. For instance, the Hubbert model of resource depletion has been described by AB models (Bardi 2005) and using system dynamics (Bardi and Lavacchi 2009). Recently, an attempt of using AB methods for world modeling has been reported to be in progress (Fiddaman 2010 b).

There are other methods of modeling complex systems, just as there exist a great variety of systems that we may define as "complex," "nonlinear" or "feedback dominated." In general, the field is still very much in progress and new methods and new software are continuously being developed. As models become more sophisticated, however, the question of their applicability to the real world becomes more pressing. The next chapter will address this fundamental question and show a historical example.

Chapter 5
Modeling the Real World: Whaling in Nineteenth Century

Whaling in nineteenth century went through a spectacular cycle of growth and decline. It had started as a humble way of obtaining food for aboriginal people but, in a few decades, it became a major worldwide industry. Then, in the second half of the nineteenth century, it declined and, by the end of the century, it had practically disappeared. In this chapter, we will see how this real world system can be quantitatively described using the basic approach of system dynamics.

Before going into actual models, we need to understand how the whaling industry worked in the nineteenth century. We all probably have a mental model of those times from Herman Melville's novel "Moby Dick." It is clear from the novel that whales were hunted and killed for "whale oil" but we would have a hard time in understanding from Melville that it was used as lamp fuel.[1] For him, it was obvious, but not for us.

At the beginning of the nineteenth century, indoor lighting was still largely based on oil lamps everywhere in the world. It was an ancient technology that goes back to hunting and gathering societies which used animal fat as fuel (Kimpe et al. 2001). With the economic expansion brought by coal, population increased everywhere in the world and this created a large demand for lamp fuel. This fuel had to be liquid; coal, alone, was not practical as light source. Gas lamps, using town gas, (obtained from coal) were a possibility, but town gas was available only in large cities and only starting about mid-century. The light bulb was invented by Thomas Alva Edison in 1879 and it did not dominate the market until well in the first decades of the twentieth century. Hence, oil lamps remained in widespread use throughout most of the nineteenth century.

---

[1] Melville does provide a hint that whale oil was used to light lamps when he describes Stubb eating whale meat, saying, "That mortal man should feed upon the creature that feeds his lamp, and, like Stubb, eat him by his own light, as you may say; this seems so outlandish a thing" Chapter lxv, Moby Dick.

U. Bardi, *The Limits to Growth Revisited*, SpringerBriefs in Energy: Energy Analysis, DOI 10.1007/978-1-4419-9416-5_5, © Ugo Bardi 2011

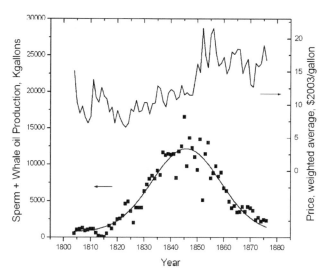

Fig. 5.1 Prices and production of whale oil in nineteenth century, from (Bardi 2007b). Reproduced by permission from (Taylor and Francis eds.)

Whales had not evolved to produce fuel for lamps used by humans, but some species stored fat in the form of a clear liquid that had the advantage of burning cleanly, leaving no odors. Whale oil turned out also to be less expensive than vegetable oils. As a consequence, in the early nineteenth century, the rapidly increasing demand for whale oil generated a whole industry. Around 1850, the American whaling industry was at its peak, producing a total that reached close to 15 million gallons of oil per year (Starbuck 1878). But decline was around the corner.

The figure above shows the production of whale oil in America (Bardi 2007b). The production data sum the contributions from the two main species captured in those times: "right" and "sperm" whales. Other species were also hunted, but their contribution was minor (Fig. 5.1).

As we see, the production curve is "bell shaped" and nearly symmetric. It can be fitted reasonably well using a Gaussian function or the derivative of a logistic curve (Bardi 2007b). This shape of the production curve is often observed for the exploitation of mineral resources, as noted by Marion King Hubbert (1956) for the case of crude oil in the 48 US lower states. Many other examples were observed of this behavior for other mineral resources (Bardi and Pagani 2007) and it is not surprising that the slowly reproducing whales behaved in the same way as non-renewable resources when exploited by a fast-growing industry. But why exactly the bell shaped curve?

According to the established views in economics, rising prices should lead to increasing production. Instead, the historical data for whale oil do not show this correlation. In the first phase of the cycle, production went up while prices remained nearly constant. Near and after the peak, prices went up but production went down. The problem, here, is that conventional wisdom in economics sees markets as basically static entities, changing only as the result of small perturbations. But the market of whale oil changed very rapidly, going from zero to 15 million gallons per year in a few decades and then back to nearly zero; all in less than a century. Indeed, here we need a different kind of approach: system dynamics which was especially designed for this kind of rapidly evolving systems.

A likely interpretation of the historical data is that the production of whale oil peaked and went down mostly because whalers were gradually running out of whales, at least of the kind that could be captured and killed by the technologies available at the time, mainly the "right whale."[2] This interpretation is confirmed by the estimations of marine biologists who have calculated that, at the end of the nineteenth century, there remained in the world's oceans only about 50 females of right whale (Baker and Clapham 2004).

Hence, whaling in the nineteenth century can be considered as an economic system dominated by the availability of the resource it utilized: whales. This is, of course, an approximation but all models are approximated. The point is to see whether a specific approximation is good enough to describe the system. Moving along this line of thought, we can see the system as a small, self contained ecology; simple enough that we can think that it can be described by a predator (whalers) and prey (whales) dynamic – that is, by the Lotka-Volterra equations. According to this line of thought, we can reduce the system to only two stocks: that of whales and that of whaling ships. Whales had few, if any, predators other than humans and whalers were specialized in hunting whales and no other species.

Now, we need to consider how these stocks vary depending on their reciprocal interaction. The Lotka-Volterra model was already described in a previous chapter: it can be written as two differential equations which couple the relationships existing between preys and predators. For the specific case of whaling, we can use an even simpler version. Whales reproduce very slowly and so we can neglect the "reproduction" term, the last one in the first equation. This is an assumption that can be confirmed by using the full model and comparing the results (Bardi and Lavacchi 2009).

So, we can write this simplified version of the model as:

$$R' = -k_1 CR$$

$$C' = k_2 CR - k_3 C$$

In this version we have two main stocks: capital ("C") and resources ("R"). "R" is (or is proportional to) the number of whales alive in the ocean while "C" is an aggregate parameter proportional to the capital accumulated by the whaling industry: ships, whalers, equipment, facilities, etc.

---

[2] Note that in the choice of the name "right whale," intended as "easy to kill," whales were not consulted.

The equations describe a situation in which the stock of whales ("R") is depleted proportionally to the number of whaling vessels and to the number of whales (k_1CR). This assumption is reasonable: the more whalers there are around, the more whales they will catch – but in order for this to happen, there must be whales and the more whales there are the more catches there will be. Then, the model assumes that the capital stock will increase in proportion to the number of catches (CR). This is a reasonable assumption which implies that the industry will grow in proportion to its profit. But the size of the industry will also decrease in proportion to its own size – this effect is often referred to as "depreciation" of the capital. This model is similar to the one known in economics theory as the "free access" model (Smith 1968).

The model can be solved qualitatively and the results are shown in the following Fig. 5.2 (obtained using the Vensim software).

As shown in the figure, the model generates a single growth cycle which appears similar to one of the cycles of the standard Lotka-Volterra model discussed in the previous chapter. However, since the resource is nonrenewable, the system never reaches a steady or oscillating "homeostatic" state. Rather, it grows and declines until it "dies" when there is no more resource available.

If the qualitative results of the model make sense, the question is whether it describes the real world's historical data. In order to verify this point, we need data on at least some of the parameters of the historical system. We do not have reliable data on the stock of whales in the nineteenth century, but we have data on the production of whale oil in the United States (Starbuck 1878) and that is something that we can use as proportional to the "production" (R') parameter. Regarding the "Capital" stock, we may use a "proxy" set of data, assuming that it is proportional to the number (or total tonnage) of the available whaling ships in America; another known set of data (Starbuck 1878).

At this point, the constants in the model can be varied in order to fit the experimental data and it turns out that the model can describe the historical evolution of the system; as we can see in the figure (Bardi and Lavacchi 2009) (Fig. 5.3).

The fact that the model can fit the data does not mean that it is the only one that can do so, or that it may be considered automatically validated. However, fitting this kind of curves is a difficult test: it is reasonably easy to fit a single parameter (production or capital build-up), but not easy at all to find a set of constants that can fit both. In this sense, the ability of the model to describe the data is a strong indication that the two stocks of the model, resources, and capital, are indeed related to each other in a feedback relationship.

One element to be discussed about whaling is the possibility that the decline in production, may have been caused by technological, rather than economic, factors. Could it be that whale oil was being replaced by a cheaper fuel, kerosene, derived from crude oil? That was surely a factor in the final stage of the whaling cycle but, at the time of the first oil well dug in Pennsylvania, in 1859, the production of whale oil was already half of what it had been at the peak. It took several more years for kerosene to overtake whale oil (Laherrere 2004). Besides, kerosene was widely considered an inferior fuel to whale oil and it is likely that only the ongoing decline of whale oil production – and the high associated prices – stimulated the growth of kerosene production. So, the model based on resource depletion remains valid for most of the productive cycle.

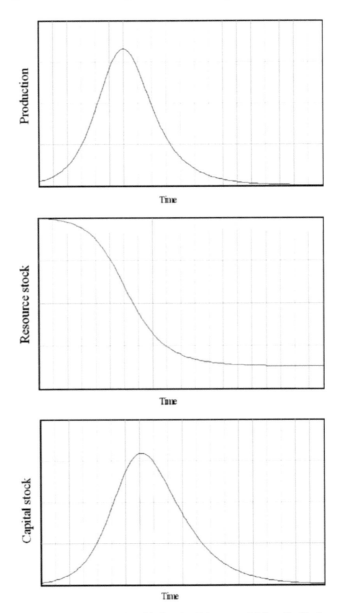

Fig. 5.2 Qualitative solutions of the simplified Lotka-Volterra model described in the text. Results obtained using the Vensim software

Of course, the model neglects several important factors such as prices and technological progress. Still, it can be seen as a "zero order" model that manages to catch the main elements of the history of the American whaling industry. According to the model, we have an example of an economic cycle generated by the progressive

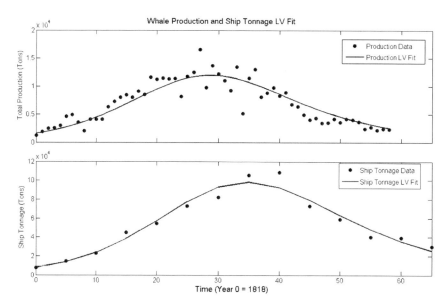

Fig. 5.3 Quantitative fitting of the historical data for whale oil production. *Upper curve*: oil production; *lower curve*: ship tonnage. From (Bardi and Lavacchi 2009). Image credit: MDPI

overexploitation of a slowly renewable resource. The industry grew on the profits of its initially abundant resource. In time, however, the dwindling number of whales reduced the industry's profits. At some point, these profits became too low to maintain the industry's growth and, eventually, caused the start of its decline. This decline was probably accelerated in later times by the development of alternative fuels such as kerosene.

Other historical cases where the model was found to be able to describe real economic systems are those of gold extraction in California and in South Africa and oil production in the US 48 lower states and in Norway (Bardi and Lavacchi 2009). It is likely that there are many more systems that can be modeled in this way.

This approach tells us that even very simple models based on dynamical concepts can provide a good fit with historical data and illustrate how negative and positive feedbacks combine in the process of resource exploitation. Obviously, system dynamics is a vast field and the model described here is just a starting point, so simple that some system scientists would not recognize it as part of the "system dynamics" family. However, simple models such as this one have the advantage that a quantitative fitting of the experimental data is possible; an approach normally not possible and often not even sought for, in more complex system dynamics models. But the Lotka-Volterra model was shown here just a starting point for understanding the basic elements of system dynamics. More complex models will be examined in detail in the following chapters.

Chapter 6
World Modeling by System Dynamics

In the history of science, bridging the gap between different scientific fields usually leads to new insights and new discoveries, as it has happened many times from the age of Galileo. In the 1950s, one such gap was bridged when there started to appear the concept that some mathematical models used in engineering could be used for other kinds of systems, such as social and economics ones.

It was already known that some mechanical systems were highly nonlinear, such as the governor valve of a steam engine. These systems showed the negative and positive feedback features that were typical of economic systems, and also of some social and biological systems. With the development of digital computers, the ambitious idea of using methods developed in engineering for describing economic and social systems seemed to be very promising. That opened the way, eventually, to "The Limits to Growth" (LTG) study, which attempted to model the whole world in its economic, ecological, and social aspects.

Arnold Tustin, of the London school of economics, was one of the first to work with non linear models in economics with his book titled, "The Mechanism of Economic Systems: An approach to the problem of economic stabilization from the point of view of control system engineering" (Tustin 1954). Another engineer, R.G.D. Allen published "The Engineer's Approach to Economic Models" (Allen 1955). Alban Phillips – an economist – published some papers on the subject of using engineering methods for modeling the economy (Phillips 1950, 1954).

All these studies recognized the importance of the same elements: time delays, feedback loops, stocks, flows, and the fact that dynamic systems are not explainable by simple cause and effect relationships. It was a very innovative set of ideas that could be put into practice only by means of the newly developed digital computers. With computers, for the first time in history, it was possible to examine the behavior of systems of many coupled differential equations quickly and easily.

A fundamental contribution in this field came from Jay Forrester who, in the 1960s at the Sloan School of Management of MIT, had started developing his socioeconomic

U. Bardi, *The Limits to Growth Revisited*, SpringerBriefs in Energy: Energy Analysis, DOI 10.1007/978-1-4419-9416-5_6, © Ugo Bardi 2011

dynamic models at the level of cities and neighborhoods. He called these studies "urban dynamics" in a book with the same title (Forrester 1969). Later, the concepts of "Industrial Dynamics" and of "Urban dynamics" were merged in the more general term of "System Dynamics" which, in Forrester's view, described the application of the method to all kinds of economic and social systems. Here is how Forrester explained the way he saw this field (Forrester 1998).

> People are reluctant to believe physical systems and human systems are of the same kind. Although social systems are more complex than physical systems, they belong to the same class of high-order, nonlinear, feedback systems as do physical systems. The idea of a social system implies that relationships between its parts strongly influence human behavior. A social system strongly confines behavior of individual people. In other words, the concept of a system contradicts the belief that people are entirely free agents. Instead, people are substantially responsive to their changing surroundings. To put the matter more bluntly, a social system implies that people act partly as cogs in a social and economic machine. People play their roles while driven by pressures from the whole system. Accepting the dominance of social systems over individuals is contrary to our cherished illusion that people freely make their own decisions.

In time, Forrester used the concepts of this new field to study the whole world's economic system. In 1971, he published his work on world modeling with the title "World dynamics" (Forrester 1971).

Forrester's book was not meant as an exhaustive study. The task of modeling the world in a more detailed way fell to another study, the Limits to Growth, a work promoted by the Club of Rome and performed by a group of young scientists (Meadows et al. 1972). Technical details on this study were published in a later work (Meadows et al. 1974). Later, in 1992 and in 2004, some of the authors of the first LTG study published updates of their calculations under the titles of "Beyond the Limits" (Meadows et al. 1992) and "Limits to growth, the 30 years update" (Meadows et al. 2004). The models and the database were refined and expanded in these later studies but the results were, basically, the same.

The models underlying "World Dynamics" (Forrester 1971) and of "The Limits to Growth" (Meadows et al. 1972) are based on similar principles. The world is divided into a number of subsystems; agriculture, industry, population, etc., and the relations among the elements of the subsystems are described by a set of equations which are solved iteratively in order to determine the evolution in time of the system.

The art of modeling lies in selecting the elements of the system. A model which integrates most of the subsystems together in a small number of parameters is said to be "aggregated." The more elements are aggregated together, the more the modeling will be approximate, but an aggregated model requires a smaller number of assumptions and is, therefore, less sensitive to the inherent uncertainties that are unavoidable when estimating a large number of input parameters.

Forrester and the LTG team both chose highly aggregated models, something for which they were criticized later. But, with these early models, it was the only possible choice and a good one as long as its limitations and its purpose are understood.

Fig. 6.1 A simplified representation of the five main elements of the world3 model: population, pollution, capital, agricultural resources, and mineral resources. From Myrtveit (2005), image courtesy of Magne Myrtveit

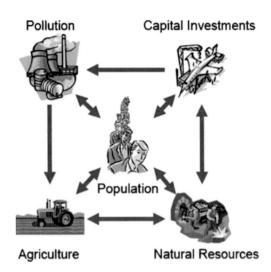

Here, we'll consider in detail only the LTG study. In the first published book (1972), the model was already very complex, with hundreds of parameters and many levels. Nevertheless, the model emphasized the interaction among five main sectors which may be summarized as,

1. Human population
2. Non renewable resources (minerals)
3. Renewable resources (agriculture)
4. Capital resources
5. Pollution

The five elements are shown in a schematic graphical form (Myrtveit 2005) as (Fig. 6.1):

This scheme has to be taken with caution as it is a considerable simplification of the actual model. However, it is a useful way to gain some confidence with the main features of the model. In practice, the graphical version of the complete world model, with all the interactions among elements, takes up two full pages of the first LTG book.

The behavior of the subsystems of the model can often be described in relatively simple ways. For instance, the casual loop diagram shown below can describe the industrial capital subsystem (Fig. 6.2).

These interactions can also be described in words, rather than as equations or diagrams. As an example, the paragraph below describes the "core" of the World3 model in terms of the interactions of the industrial capital element with the other elements of the model (from the 2004 edition of LTG (Meadows et al. 2004)):

The industrial capital stock grows to a level that requires an enormous input of resources. In the very process of that growth it depletes a large fraction of the resources available. As resource prices rise and mines are depleted, more and more capital must be used for

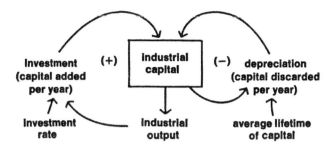

Fig. 6.2 Causal loop diagram of the Capital sector of the world3 model From Meadows et al. (2004), image courtesy of Dennis Meadows

> obtaining resources, leaving less to be invested for future growth. Finally, investment cannot keep up with depreciation, and the industrial base collapses, taking with it the service and agricultural systems, which have become dependent on industrial inputs.

As it can be noticed, capital and resources behave in the model in a manner similar to how "predators" and "prey" do in the Lotka-Volterra model that was described in the previous chapter. In other words, "Capital" (an aggregate of economic resources) grows on natural resources as a predator grows on its prey. Of course, the simple predator–prey relationship is embedded in a much more complex world model where a large number of elements interact with each other. Then, the relations among the elements are not always as simple as in a "predator–prey" model. For instance, here is the loop for the "population" section (Fig. 6.3).

As implemented in the world3 model, this subsystem considers several factors including the fact that population growth is not simply proportional to the availability of services and food. This is a well-known phenomenon: the availability of such services as instruction, the existence of social equality and – in general – of wealth, reduces the growth rate of human population. It is a phenomenon known as "demographic transition." It was discovered for the first time in 1929 by Warren Thompson (Encyclopedia of population 2003).

The general behavior of the world system described by the world3 model turns out to be characterized by robust features. Most scenarios show that, at the beginning of the simulation, the positive feedbacks associated with the exploitation of abundant natural resources lead agricultural and industrial production to grow rapidly, while pollution does not pose constraints on growth. In these conditions, society accumulates industrial capital in the form of equipment, resources, and human knowledge. With time, however, natural resources become increasingly depleted and more expensive to produce. At the same time, pollution increases and becomes a significant negative factor affecting growth. As time goes by, more and more capital is needed to maintain growth and, eventually, it becomes impossible to keep agricultural and industrial production growing; both reach a peak and start declining. Afterwards, the accumulated capital also shows a peak and starts to decline. Together with these phenomena, population also follows a cycle: it grows with the increasing production of food and services. Population growth affects the availability of capital, increases the depletion of natural resources, and

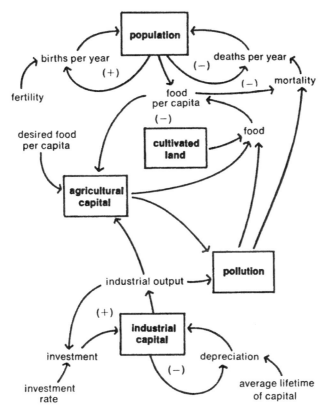

Fig. 6.3 Causal loop diagram of the population sector of the LTG model from Meadows et al. (2004), image courtesy of Dennis Meadows

generates pollution. It stops growing some years after the peak of industrial and agricultural production, as a result of the reduced food and services available.

This behavior is most evident in the simulation that the authors call the "standard run" or the "base case" model. In all three versions of LTG, the base case run is the one that used as input the data that were, according to the authors, the most reliable available. In Fig. 6.4, we see a comparison of the results obtained in 1972, with the first version of LTG, and those shown in the most recent version (Meadows et al. 2004).

Note that the vertical scales of the two figures are not the same; apart from this point, however, the overall results are very similar. The peaking of the world's industrial and agricultural production occurs at about the same time in both runs, while the human population curve shows an earlier peak in the most recent calculations. This difference is the effect of using different assumptions in the way the population stock is related to the other elements of the model.

In the base case scenario, pollution is not the most important constraint to growth. Rather, the main factor in slowing down growth is the need to divert more and more capital resources to maintain the production of both food and services. Since capital

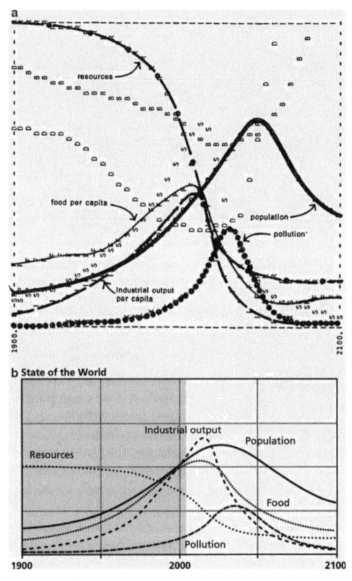

Fig. 6.4 Comparison of the "Base Case Model", calculation from the 1972 and 2004, Editions of "The Limits to Growth" (notice the difference in scale: the vertical scale of the 2004 run is twice as large as that of the 1972 run). From Meadows et al. (1972, 2004). Image courtesy of Dennis Meadows

resources are limited, the overall result is that growth slows down; production peaks, and then declines. In a way, this behavior is the result of forces which act in a similar manner as those which brought down the whaling industry in the nineteenth century, as it was described in the previous chapter.

The analyses given independently by Turner (2008) and by Hall and Day (2009) show a remarkably good correspondence between the behavior calculated in the LTG "base case" scenario and the historical data up to the present time. Hall and Day (2009) also make the point that it was simply the primitive state of drawing *x* axes with the computer technology of 1972 that seems to have misinformed, or allowed, the critics to conclude that the model was incorrect.

It is not possible, yet, to say whether the collapse envisaged in the "standard run" of LTG is starting to take place right now, as may be indicated by the ongoing recession, the apparent cessation of rapid increase in oil production, and the increasing food uncertainties in 2011. If we do see the economic system continuing to decline in the near future, the 1972 "base case" model would turn out to have had an unexpectedly good predictive value.

But prediction is not, and never was, the main purpose of LTG. It was, rather, to give to users a mental model of how changing the input for the various elements changes the output of the model. The model and its results can't be expected to predict the future but are to be seen as giving us a way to be prepared for it.

Indeed, LTG was intended from the beginning to provide a fan of scenarios which would examine and evaluate various "what if" hypotheses. What if natural resources are more abundant than we think? What if there is a technological breakthrough in energy production? What if humans agree to stop population growth? This and many other effects can be tested using the world3 model developed for LTG.

Over three different books (Meadows et al. 1972, 1992, 2004), the authors undertook a major effort to explore and describe a wide range of scenarios. Here, it would be impossible to go into all the details; the reader is invited to examine the latest version of the LTG study (Meadows et al. 2004) for a summary of the work done. However, some important points can be at least summarized here.

Basically, the LTG study examined three different sets of possibilities:

1. Different parameters in input: e.g., larger amounts of natural resources.
2. Technological effects: e.g., higher efficiency in resource exploitation, higher agricultural yields, and better ability to abate pollution.
3. Policy changes: e.g., capping the rate of resource exploitation or population control.

The first set of possibilities, different parameters in input, deals mainly with the exploration of the hypothesis "What if nonrenewable resources are more abundant than assumed in the base case scenario?" However, larger initial stocks of resources do not change the basic behavior of the model. With more natural resources, industrial and agricultural production grow to higher levels than in the "standard run," but collapse takes place anyway, although later in time, being generated either by depletion or by the effect of unchecked pollution (or both effects at the same time).

The model can also to deal with technological innovation, although somewhat indirectly. The model does not have an explicit "progress" parameter, but it is possible to model innovation in terms of its effects on other parameters. For instance, a breakthrough in energy production would manifest itself in the availability of more

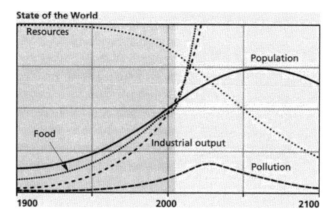

State of the World

Fig. 6.5 The "IFI-IFO" scenario of LTG; a simulation that removes all physical limits assuming that resources are infinite, just as is the human capability of fighting pollution. From Meadows et al. (2004), image courtesy of Dennis Meadows

abundant nonrenewable resources, since with more energy we would be able to exploit lower grade mineral ores (Bardi 2008a). Other breakthroughs might involve more efficient agricultural techniques that would appear in the model as a higher yield in food production. These and other possibilities, however, do not qualitatively change the results of the simulations. Collapse still arrives, although later in comparison to the base case scenario.

Even for the case of "infinite mineral resources" (simulated as an ever-increasing technological efficiency in recovering them), collapse is the eventual result of uncontrolled pollution and the lowering of agricultural yields.

Not one, but several technological breakthroughs are needed in order to generate a scenario in which growth of both the economy and of population continues unabated for the whole of the twenty-first century. This is the scenario dubbed IFI-IFO by the LTG authors to mean "Infinity In, Infinity Out." This scenario assumes infinite nonrenewable resource, infinite ability of agriculture to produce food, and infinite ability to abate pollution. The results are in Fig. 6.5.

This run is not realistic, obviously, but it may be useful to show how the model behaves in these assumptions. As mentioned earlier on, the human population does not simply grow in proportion to the availability of food. The "demographic transition" puts a limit to the number of humans on this planet even in these very optimistic assumptions on resource availability and on pollution control.

All these scenarios describe the results of a system where the economy operates according to "business as usual" assumptions. That is, it is assumed that people will continue to exploit natural resources at the fastest possible speed and that they will reproduce according to the same behavior they have shown in the past. However, these assumptions can be changed. This is equivalent to assuming that governments intervene, or that social mores change to encourage birth control in order to reduce population growth, or to curb resource exploitation.

The effects of population control were studied already in the first edition of the LTG of 1972. The simplest assumption was that population could be kept constant at the levels of 1972. Clearly it was an unrealistic assumption; nevertheless it did not prevent collapse of the industrial and agricultural systems. Again, these results show how robust the behavior of the model is. The results of the most recent LTG study (Meadows et al. 2004) are essentially the same as those of 1972 in that they show that simply keeping population constant – even if it were possible – would not prevent economic collapse.

Nevertheless, it is possible to adjust the parameters of the model in such a way as to obtain a stable state that does not imply a loss of material standards (or only a small one) for human beings, at least on the average. In order to reach this state, it is necessary to curb the growth of both population and industrial output; together with attaining considerable technological advances in fighting pollution and reducing the problem of the erosion of agricultural land. In other words, stabilization is possible only by stabilizing the whole economy and not just one of its elements such as population. So, this scenario is related to the concept of "steady state economy" as expressed, for instance, by Herman Daly (1977) (Fig. 6.6).

The fact that the LTG authors tried to find parameters that would lead to a steady state economic system generated much controversy. Advocates of continuing growth saw the idea of a steady state economic system as impossible or counterproductive or both. Georgescu Roegen, well known for his attempt of introducing thermodynamics into standard economics theory (Georgescu-Roegen 1971) did not believe that the economic system could be stabilized (Georgescu-Roegen 1977, 1979a, b). Even recently, for instance, Gail Tilverberg (2011) argued that the world's financial system is based on debt and that the only way to repay it is to continue growth.

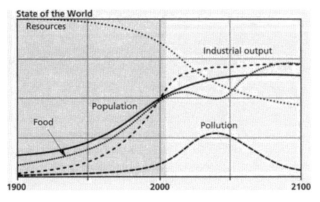

Fig. 6.6 The "Stabilization" scenario of LTG. A simulation that assumes a concerted intervention to control population and curb resource exploitation. With the help of reasonable technological progress, these assumptions lead to a scenario where collapse of the world's economy does not occur within the twenty-first century. From Meadows et al. (2004), image courtesy of Dennis Meadows

Nevertheless, it is also true that complex systems that receive a flux of energy from an external source do tend to reach a steady state or, at least, oscillate around a constant state; a condition called homeostasis (Lotka 1925; Karnani and Annila 2009). The world system studied by LTG is no exception; the calculations reported show a tendency to stabilization around the end of the twenty-first century that perhaps would lead to a steady state in the twenty-second century corresponding to an agricultural society at a level possibly comparable to that of the eighteenth or the seventeenth century. But the search undertaken by the authors of LTG in 1972 was not about that kind of steady state. It was about a steady state at a level of industrial and agricultural production, as well as of population, comparable to the present one. Obviously, the models indicated that the world system, left to itself, would not attain such a state. That led the authors to propose to tweak the parameters of the system in such a way to force it to reach the desired state. This approach involved drastic changes in the way the economic system works; something that would probably require heavy government intervention or – anyway – a considerable change in the way decisions are taken in the world. That was one of the elements that, later on, made the LTG authors unpopular.

In any case, finding such conditions was possible for the three dates for which the simulations were performed (1972, 1992, 2004) but it became increasingly difficult as the world economic system gradually ran out of resources and persistent pollution became more important. At the larger population and increased depletion level of today (2011), it may no longer be possible to find a new set of constraints that would stabilize the system at the present level.

Independently of whether stabilization is possible or not, the very fact that the LTG authors looked for such an outcome implies a judgment on what future is good for our society. The judgment of the LTG authors and of their sponsors, the Club of Rome, was that stabilization is a good thing to obtain. Others (e.g., Boyd 1972; Vajk 1976) have used the world3 model in order to find a different future that they judged desirable, that is how it could be possible to keep the economy growing forever. This is a perfectly legitimate use of the model; although based on a different evaluation of what the human future should be.

The two reports by Forrester (1971) and by the LTG group (Meadows et al. 1972) were not the only ones dedicated to world modeling at that time. Several other groups engaged in the same task using similar or different methods. Some attempts at world modeling appeared even before the first LTG. An example is the book by John Harte and Robert Socolow "Patient Earth" (Harte and Socolow 1971). These authors, however, did not use system dynamics. We may also cite "Overshoot" Catton (1982) which was written in the early 1970s but was published only in 1982. In his book, Catton did not use formal system dynamics methods, but the concept of "overshoot" is deeply based on dynamic concepts. Other simulations were performed in the book "Beyond Oil" by Gever et al. (1986) which generated broadly similar results to the LTG ones.

The Club of Rome directly sponsored some studies that were thought of as follow-ups of LTG. One was authored by Mihajlo Mesarovich and Eduar Pestel (1974) and was titled *Mankind at the Turning Point: The Second Report to The Club of Rome.* This work was the logical step to be performed after the first report. It attempted to expand the LTG calculations, this time disaggregating the various regions of the world which was divided in ten geographical areas. The authors found that the collapse predicted by the LTG study was to be regional first, rather than global, and that it was to arrive first in South America and Southeast Asia. Another study sponsored by the Club of Rome was performed by the Dutch economist and Nobel Prize winner Jan Tinbergen, whose work was titled, "RIO, Reshaping the International Order" (1977). However, this study emphasized economic growth rather than limits and, eventually, the Club of Rome withdrew its support of it.

Other studies and documents based on world modeling were not sponsored directly by the Club of Rome, but were inspired by the LTG study. Several such models are described in the book "Groping in the dark" (Meadows et al. 1982). Among these studies, for instance, in 1977, Amilcar Herrera, of the Bariloche Foundation, Argentina, used system dynamics to study the future of the Latin American economy (Herrera 1977). This work attempted to refute the predictions of collapse of the LTG study but, despite some very optimistic assumptions in input, developing convincing scenarios of this kind was not possible. We may also mention the *Global 2000 Report* for President Carter written by a team of authors under the direction of Gerald O. Barney (1980). This was the first (and, so far, the last) report commissioned by a major national government on the economic, demographic, resource, and environmental future of the world. The results were broadly consistent with those of LTG but, despite selling over 1.5 million copies in eight languages, this study left little trace in the debate.

On the whole, the effort in world modeling that had started in the 1960s faltered in the late 1970s and in the 1980s for lack of resources and as a consequence of the growing negative opinion on LTG. The LTG authors kept working on their models and published updated version of their world simulations in 1992 and in 2004. However, world modeling was largely abandoned for more than two decades and, 40 years after the publication of the first LTG we are facing the future without having planned for it.

Chapter 7
Criticism to "The Limits to Growth"

Thomas Huxley (who liked to be defined "Darwin's bulldog") said that "It is the customary fate of new truths to begin as heresies and to end as superstitions." There are many cases in which excessive conservatism in science has prevented new ideas ("heresies") from being adopted and has kept old ideas ("superstitions") alive for too long. Conservatism, however, is not a negative feature of science. It is useful, actually essential, in order to maintain intact the core methods of science and eliminate misplaced enthusiasm not based on sufficient evidence. The problem is: what is a genuine innovation and what a fad that will soon disappear?

In practice, a new scientific theory must expect to pass a very tight screening before it is adopted. The screening will be the tighter the more radically innovative the new theory is and it will be especially harsh if the proposers do not belong to the recognized elite in the field or, worse, are outsiders coming from a different scientific field. In some cases, it is well known that a new theory can get a foothold in science only when its original opponents retire and a new generation of scientists takes over.

In the case of the application of system dynamics to social and economics systems, the work presented by Forrester (1971) and by the LTG group (Meadows et al. 1972) had all the features of a radical innovation. In addition, it was proposed by people who had no pedigree in the field of economics or in social studies. As one might imagine, the resulting debate turned out to be harsh, to say the least.

After the publication of LTG in 1972, comments arrived from different fields of science and many were favorable, actually enthusiastic. People were impressed by the width and depth of the studies, by the innovative approach, and by the use of computers; at that time a novelty. As an example, Robert Townsend, author of "Up the Organization" (1970) wrote about "The Limits to Growth" (cited in Simmons 2000).

> The *Limits to Growth* has made headlines the world over. Its shock waves have caused our most cherished assumptions to come crashing down. It is a book that we can ignore only at our peril.

> If this book does not blow everybody's mind who can read without moving his lips, then the earth is kaput.

U. Bardi, *The Limits to Growth Revisited*, SpringerBriefs in Energy: Energy Analysis, DOI 10.1007/978-1-4419-9416-5_7, © Ugo Bardi 2011

These first enthusiastic reactions were rarely based on an in-depth examination of the study since very few people, at the beginning, had the technical ability and the means to replicate the work of Forrester and of the LTG team. Nevertheless, many researchers tried to adapt the new methods to their own field of study and to examine more in detail the subject of world modeling. A good example of this positive approach is the 1973 paper by John Bongaarts (1973), a demographer who looked at the population subsystem of the world3 model. There are other examples of positive reaction to LTG, even among economists, for instance the article by Brown et al. (1973) in the "Oxford Economic Papers."

So, the idea that LTG was immediately laughed off as an obviously flawed study is just one of the many erroneous legends that surround it today. Nevertheless, it is also true that the first appearance of dynamical world modeling generated a strong negative reaction, especially from economists. The present chapter is dedicated to examining the criticism to LTG which was based (or claimed to be based) on scientific arguments. The more specifically "political" reaction to the study will be examined in another chapter.

Three economists, Peter Passel, Marc Roberts, and Leonard Ross, fired the first broadside against the LTG study on 2 April 1972 in an article that appeared on the New York Times. The academic qualifications of the authors and the large diffusion of the magazine gave to the article a considerable weight.

Looking today at the article by Passel and the others, we see that it was very limited. In two pages of text, the authors could not do more than attack LTG by means of a series of declarations of disbelief, mainly focused on the data in input that were defined as "too pessimistic." Much of this criticism was generated by a misunderstanding of the meaning of one of the tables of the second chapter of the LTG book which described the expected duration of some mineral resources in various assumptions. Passel and the others took it for a "prediction" whereas it should have been clear from the text that it was just an illustration of possible trends in some specific assumptions which the authors themselves defined as "nonrealistic." In later years, this wrong interpretation would be picked up again (Bailey 1989) and used for a political attack against the LTG study but, in this phase, it passed almost unnoticed.

The reaction of Passel and the others is typical of the initial "gut reaction" to LTG of many economists. Already in 1971, Martin Shubik (1971) had commented "World Dynamics" by Forrester (1971) in "Science" saying, "What is this book for? Its behavioral-scientific content is virtually zero." Wilfred Beckerman (1974) blasted at LTG in an article that appeared in "Oxford Economic Papers" saying that it was a "brazen, impudent, piece of nonsense." Yale economist Henry C. Wallich labeled the book "a piece of irresponsible nonsense" in his 13 March 1972 Newsweek editorial. John Koheler wrote in the "Journal of Politics" (Koehler 1973) that, "If the point of this book is simply to observe that as t becomes large with the passage of time, ae^t becomes large, then some significant portion of its 205 pages are unnecessary."

This early criticism has only a modest interest for us, today, except as an indication of the hostile reaction of economists to the LTG approach. However, with time, more people became familiar with the theory behind LTG and this led to more serious attempts of a critical appraisal of the study. Two early works are especially important as they had a considerable impact on the debate. One was prepared by the "Sussex Group," also known as the "Science Policy Research Unit" (SPRU) of the University of Sussex in England and the other by William Nordhaus, at Yale University.

We will examine first the work of the Sussex Group (Cole et al. 1973) which remains today one of the most serious attempts ever made of a complete evaluation of the LTG model. The SPRU group of the University of Sussex was an interdisciplinary team that included not only economists, but physicists, mathematicians, historians, and other specialists. Their study on LTG appeared in 1973 in Great Britain with the title "Thinking about the Future" and, in the US, with the more imaginative title of "Models of Doom." The American edition of the book left a chapter to the LTG group to respond to the criticism, but the British edition did not. Not giving the criticized authors a chance to respond to the criticism is surely a break of the accepted rules for this kind of intellectual confrontation. We have, here, one of the many indications that the debate on LTG was harsher than the usual scientific debate.

"Models of Doom" contains a total of 14 essays, 9 of which are dedicated to a technical assessment of LTG for a total of 134 pages written by nine authors. It is impossible to discuss all the details of this study here, but the main points made in the book can be summarized by saying that criticism against LTG was based on two issues: the first was that the authors had been too pessimistic in their assumptions on the parameters in input to the model; the second that there were fundamental faults in the model used.

For the first point, pessimistic assumptions, many of the statements we can read in "Models of Doom" look simply naive today. A good example is the article on agricultural resources by Pauline Marstrand and K. R. L. Pavitt. Here, the authors suggest that the remedy to a possible world shortage of food can be obtained by cultivating the Australian continental land, which they claim to be "underutilized." The concept seems to be – all what is needed to do in order to transform the Australian desert into fertile land is to cut away the shrubs, spray pesticides, and add fertilizers and irrigation. If nothing else, the devastation caused to conventional Australian agriculture in 2010–2011 from drought and floods should put that idea to rest. But for a detailed description on how some dreams of easy abundance can be shattered, it is worth reading the chapter on Australia in "Collapse" by Jared Diamond (2005). The Australian land has been badly damaged by government policies that supported the destruction of native vegetation to be replaced by crops. Salinization and land degradation was the result. Phenomena of land degradation caused by human action occur everywhere in the world (Ponting 2007).

In "Models of Doom," we can find a similar overoptimistic approach applied to mineral resources. In the chapter by William Page, dedicated to the nonrenewable resources subsystem, we read that the earth's crust is "25–40 miles deep" and the author seems to seriously believe that minerals can be extracted from such depths. But, still today, the deepest holes ever made into the ground, oil wells, are no more than about 10 km (6–7 miles) deep and normally much less than that. Mines do not reach such depths, most of the times not even remotely. In any cases, exploitable mineral deposits ("ores") normally are the result of hydrothermal processes which occur at the surface or at low depths. Digging deeper is not a solution for mineral scarcity.

Despite these obvious shortcomings, the discussion about resources in "Models of Doom" is not completely outdated. For instance, the chapter on energy resources by A.J. Surrey and A. J. Broomley, is worth reading for its accurate evaluation of the existing reserves of fossil fuels. However, there is a basic problem with this section: the calculations reported in the LTG study took into account a wide range of assumptions, some so optimistic that it can only be described as nonrealistic. So, saying that mineral or agricultural resources may be more abundant than assumed in the "base case" model is, at best, a marginal criticism of LTG and it does not affect the validity of the study.

Hence, the other point on which the SPRU group mounted a criticism of LTG is more interesting: it is the validity of the model itself. Here, the work of the two authors engaged in this analysis, H.S.D. Cole, a physicist and R. Curnow, a mathematician, remains today one of the most detailed critical examinations of the world3 (Meadows et al. 1972) and world2 (Forrester 1971) models.

Despite the qualifications and the experience of the authors, some comments by Cole and Curnow highlight how difficult it was (and it still is) for people not trained in system dynamics to understand the aim and the structure of a dynamic model. For instance, they say, "But if he (Forrester) had started his model run in 1880 with initial values based on his arguments in World Dynamics the collapse predicted by the standard run would be brought forward by 20 years. And if he had started the model in 1850 the collapse would be predicted for 1970." Obviously, as noted in the answer of the LTG authors in the American edition, if they had chosen the start date for their simulations as – say – 1850, they would have selected values for the constants of the model such as to adapt to the known historical data. So, the collapse would still have occurred for the same date in the simulations.

The same misperception of the structure of a dynamic model can be seen when Cole and Curnow state, "In the world3 model, it is important to note that the effect of increased expenditure on anti-pollution equipment is to raise the capital use and to increase pollution." That is true but it would be a criticism only if it were intended to mean that the *only* effect of making antipollution equipment in the model is to increase pollution. That is not, obviously, the case: in the model, the manufacturing of antipollution equipment must be considered *also* as an element decreasing pollution. It would not be surprising, anyway, that something designed to fight pollution were to turn out in practice to create more of it than it can remove.

Despite these misunderstandings, the in-depth examination made by Cole and Curnow on the world3 model is still worth reading today. For instance, they observed that the world3 model had troubles in working backwards in time. They found that – going backwards – the model produced a sharp growth in industrial production around 1880. Seen in the normal direction of movement of time, that would have appeared as a decline; actually a catastrophe that, obviously, did not take place in real history.

Cole and Curnow had identified for the first time a general problem of dynamic models, the capability of working backwards ("backcasting"). It is a problem that has been examined in a number of later studies (see, e.g., Coyle 1983). It is normally believed today that a good model should work both forwards and backwards, but it is also clear that this is not a crucial property. The world 3 model had not been devised to work backwards and hence it is not surprising that it could not be made to do so. The point is whether this behavior can negatively affect the results of the calculations. Here, however, Cole and Curnow were unable to demonstrate that backcasting instabilities in World3 generated significant problems when the model was used in the normal way.

So, in the end, what do Cole and Curnow conclude? Overall, their main criticism, repeated over and over, is related to the value of the model for an actual representation of the real world. In other words, Cole and Curnow discuss how it is possible to validate the LTG model and maintain that such a validation is not possible; hence the model, in their opinion, is flawed. But this conclusion depends on what is exactly intended for "validation."

For a physicist, validating a model consists in fitting the calculated results to the experimental data. Repeated tests on different sets of data are normally performed. If the fitting is deemed to be statistically significant, then the model is considered validated for the systems being considered. This is possible, however, mainly for linear models. For instance, Newton's law of gravity can be easily validated by measuring the movement of a body in a gravitational field. Once this is done for a large number of systems, the validity of the model is not in question any more.

The problem with complex system such as a socioeconomic ones is that the behavior of such systems is, in general, not repeatable and that, often, the data available are limited. In some cases, it is possible to show that simple dynamical models describe historical systems reasonably well, as it was shown in a previous chapter for the case of whaling (Bardi and Lavacchi 2009). That, however, does not prove that a different and much more complex dynamical model can describe the world's economy throughout the twenty-first century – especially if the model is run in the twentieth century.

A physicist will often tend to evaluate all models as if they were describing linear, repeatable systems. Facing a dynamical model of a nonlinear and nonrepeatable system which is still evolving, a physicist will often consider it arbitrary, if not worthless. This appears to be the opinion of Cole and Curnow on the world3 model. But the viewpoint of the system scientist is different. John Sterman (Sterman 2002) has examined this issue (p. 521). He states:

> Because all models are wrong ([1]), we reject the notion that models can be validated in the dictionary definition sense of 'establishing truthfulness', instead focusing on creating models that are useful in the process of testing, on the ongoing comparison of the model against all data of all types and on the continual iteration between experiments with the virtual world of the model and experiments in the real world. We argue that focusing on the process of modeling rather than on the results of any particular model speeds learning and leads to better models, better policies and a greater chance of implementation and system improvement. (Sterman 2002)

---

[1] Physicists may not agree on this point. Perhaps, Sterman's statement should be modified as "all non linear models are wrong."

So, we may say that we are facing a clash of absolutes between the way of seeing the world that is typical of the physicist and that of the system scientist. Physicists seek "laws" that describe the behavior of systems and that can be used for predictions. System scientists seek for an understanding of how complex systems work.

Perhaps one can say that all models are wrong, but all models can be useful as long as one takes them for what they are: ways for understanding systems – a concept that we can define as "way of formalizing our assumptions" (Hall and Day 1977). Seen in this way, the correct question to ask about the world3 model of LTG (or Forrester's world2) is not whether or not the model can be *proven* to describe reality, but what alternative models we can conceive and how the results obtained differ. In this sense, "Models of Doom" does not provide alternatives and is therefore a weak criticism of these world models.

We see from this summary that the debate in "Models of Doom" was harsher than the normal scientific debate but, on the whole, still acceptable. The authors of the various chapters of the book expressed their opinions, they supported their statements with data, and they cited their sources. The LTG team, on their part, was given a chance to respond (at least in the American edition of the book). On this basis, one could have expected that further exchanges might have led to, at least, a definition of the areas of disagreement and, perhaps, to better models. But that did not happen. The debate, instead, degenerated into insults and political accusations, in a way similar to much of the climate debate today. In part, this was already happening in "Models of Doom" which contained a section with the ominous title "the ideological background." This section will be examined in another chapter.

We can now examine the work of William Nordhaus, who emerged out of the debate as one of the major critics of the LTG study and, in general, of system dynamics as a method for modeling economic systems. In 1973, Nordhaus published a paper titled "World dynamics: measurements without data" (Nordhaus 1973) taking as a target Forrester's book (Forrester 1971). However, it is clear that Nordhaus's attack also broadly included the LTG work.

Nordhaus's paper spans 27 pages and contains much material worth discussing, but it would be out of scope to go into all the details here. Forrester himself used 21 full pages in his response that was published in "Policy Sciences" (Forrester et al. 1974). For what we are concerned here, we may summarize Nordhaus's criticism as pertaining to basically three categories: (1) accusations ad personam, (2) unsubstantiated statements of disbelief, and (3) quantifiable criticism.

As for the first category, we can take as an example the accusation of "lack of humility," made against Forrester. The gist of this accusation is that carrying world simulations all the way to the end of the twenty-first century is much too ambitious to make sense. This is a legitimate opinion, but not something that can be evaluated on the basis of objective criteria. On this point, however, it is worth noting that Nordhaus himself, later on, committed the same intellectual fault – according to his own definition – with his DICE (dynamic integrated climate economy) model (Nordhaus 1992b).

The second category of criticism from Nordhaus, "statements of disbelief," collects alleged shortcomings of world modeling which, however, are not substantiated by actual proof. One such statement, taken as an example, in the following statement (p. 1166) where we read that, according to Nordhaus, in Forrester model:

..we discover dramatic returns to scale of the economy: if we double both the number of blast furnaces and the number of ore fields the output of pig iron quadruples

But nowhere in his paper Nordhaus demonstrates that Forrester's model produces such obviously unrealistic results. In fact, Nordhaus is simply looking at one of the several equations of the model without realizing that the output of each equation will be modified by the interaction with all the other equations and that will insure correct returns to scale. This is the essence of systems thinking: that parts interact.

Let us now consider the accusation of "measurements without data" which is the most important part of the paper and gives it its title. This is a quantifiable criticism: if it can be shown that Forrester (or the LTG group) were making models which are totally unable to describe the real world, then it is correct to dismiss their work as useless and irrelevant.

In "World Dynamics" (1971) and in "The Limits to Growth" (1972) one thing that can be immediately noticed is that historical world data do not appear in the calculated scenarios. For a reader accustomed to the common approach of "fitting" the data, that gives a bad impression. Is it possible that the authors of these studies were really so cavalier that they did not care to compare their results to real world's data?

But a more careful examination of the text of both studies shows that the authors do state that their calculations were calibrated on actual historical data. Not showing these data in the figures was a choice made in order to improve clarity. As a choice, it may be criticized, but not ignored.

On this point, note also that, in the "Models of Doom" book (Cole et al. 1973) examined before, none of the several authors engaged in the study felt that Forrester's work (or the LTG book) could be criticized in the terms used by Nordhaus. In the chapter by Cole "The Structure of the World Models" (p. 31 of Cole et al. 1973) the data used in the models are examined in detail. Some of the approximations utilized are criticized and in some cases it is said that the data are insufficient for the purposes of the model. But it is never stated that the models were "without data."

So, it is clear that the world2 (Forrester's) and world3 (models) were calibrated to the historical data – at least within some limits. On this point, although both Forrester and the LTG team made an effort to choose the parameters of the model on the basis of historical data, they also felt that their models had a heuristic rather than explicitly predictive objective. Therefore, there was no need for their scenarios to use a rigorous data fitting procedure of the type used in physical studies. Again, this is an attitude that can be criticized, but that cannot be ignored.

Forrester himself describes this attitude in his book "World Dynamics" (Forrester 1971). On page 14 (second edition) he says:

There is nothing new in the use of models to represent social systems. Everyone uses models all the time. Every person in his private life and in his community life uses models for decision making. The mental image of the world around one, carried in each individual's head, is a model. One does not have a family, a business, a city, a government, or a country in his head. He has only selected concepts and relationships that he uses to represent the real system. <..> While none of the computer models of social systems existing today can be considered as more than preliminary, many are now beginning to show the behavioral characteristics of actual systems.

System scientists have a structured approach on this point, as described, for instance, by Sterman (Sterman 2002, p. 523).

> … it is important to use proper statistical methods to estimate parameters and assess the ability of the model to replicate historical data when numerical data are available <..> Rigorous defining constructs, attempting to measure them, and using the most appropriate methods to estimate their magnitudes are important antidotes to causal empiricism, muddled formulations and the erroneous conclusions we often draw from our mental models. Ignoring numerical data or failing to use statistical tools when appropriate is sloppy and lazy

Of course, the very fact that Sterman feels that it is necessary to criticize those modelers who "fail to use statistical tools" indicates that the problem exists. Modeling socioeconomic systems using system dynamics tools is not immune to the biases that are easy to see in the ordinary political debate.

So, taking into account all this, how should we understand Nordhaus's criticism? If it is intended as meaning that system dynamics models provide only approximations of the historical behavior of the world, then it is a weak criticism that hardly justifies the statement "measurements without data." This point must have been clear to Nordhaus himself, who tried to substantiate his criticism by the following statement, referred to Forrester's world2 model:

> …..contains 43 variables connected to 22 non-linear (and several linear) relationships. *Not a single relationship or variable is drawn from actual data or empirical studies.* (emphasis in the original)

Let's analyze this sentence. First of all, Forrester's model, as all models, contains three elements: the mathematical *relationships*, or equations, the *variables* (populations, resources, etc.), and the *constants* which appear in the equations and which determine the quantitative behavior of the model. Nordhaus speaks here only of two of these elements: variables and relationships, but not of the third; the constants. Clearly, he was aware that Forrester was using constants derived from real world's data. But, then, what does it mean that, "Not a single relationship or variable is drawn from actual data or empirical studies"?

Evidently, Nordhaus thinks that the equations and the variables of the model should have been determined by fitting the experimental data. This is an approach that often goes under the name of "econometrics." This term does not describe a specific type of model, but it refers to a series of methods and techniques used to fit a set of data, typically a time series, to a model (Franses 2002). Econometrics can be used to test a model but, in some cases, it is the "best fit" of several models that determines which one is to be chosen. This is a legitimate technique, but one that may easily lead the modeler astray if the physical elements of the system are not sufficiently understood.

In any case, the "best fit procedure" tells you little about the physics of the system being studied. Think of Newton's law of universal gravitation. The scientists who worked before Newton on planetary motions, from Ptolemy to Johannes Kepler, had basically used a "data fitting" procedure to describe their observations but never could derive the law of universal gravitation using that approach. Instead, Newton devised a law that he thought plausible. Maybe he got the idea watching an apple falling from a tree, but that hardly qualifies as data fitting. Then, he calculated the motion of the planets according to his law. He found that simulated bodies orbiting around the Sun would describe elliptical orbits, just as it was observed for the planets. At this point, he could vary the "g" constant in his law in such a way that it was possible to use the equation to describe the movement of real planets.

So, if Nordhaus' criticism to Forrester were to be applied to Newton's gravitation law, then one should criticize it because it is not "drawn from actual data or empirical studies" One could actually criticize Newton for performing "measurements without data."

Of course, Forrester's model is much more approximate and tentative than Newton's law of universal gravitation. Nevertheless, the considerations about the validation of the model remain valid. So, in order to prove his point, "Measurements without data," Nordhaus needs to do more. He needs to demonstrate that Forrester's model is totally unable to describe reality.

So, Nordhaus sets up in his paper to "evaluate the specific assumption in the subsectors of World Dynamics" (p. 1160). The examination of the population subsystem is crucial in this analysis. In Fig. 3 of his article, Nordhaus plots data on the birth rate as a function of the Gross National Product for several countries, together with what he claims to be the results produced by Forrester's model (Fig. 7.1).

From this figure, it would seem that Forrester's assumptions are completely wrong and this is, indeed, Nordhaus's conclusion. But what is the curve that Nordhaus calls "Forrester's assumption"? In the article, we read that this curve is "Forrester's assumed response of population to rising per capita non food consumption *when population density, pollution, and per capita food consumption is held constant*" (emphasis added).

But this is *not* Forrester's assumption. Nordhaus had simply taken one of the equations from Forrester's model and had plotted it keeping constant all parameters except one (the "non food consumption" that he equates to GNP). But Forrester's model was never meant to work in this way.

This point deserves to be explained in more detail. Let us consider a simple Lotka-Volterra model, as it was described in a previous chapter, where the two stocks are foxes and rabbits. Using Nordhaus's approach for this model would mean to take one of the two equations, say the one that describes the flow of rabbits, and see how it behaves when "all the other variables are kept constant"; in this case the number of foxes. Here is the equation, where "R" is the number of rabbits and "C" is the number of foxes,

$$R' = -k_1 CR + k_2 R.$$

If C is a constant, the equation simplifies to,

$$R' = kR.$$

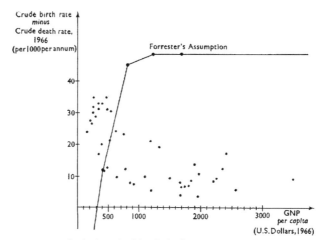

Fɪɢ. 2.   Assumed and Cross Section Population Growth, 1966.

Fig. 7.1 This figure, taken from Nordhaus (1973) attempts to demonstrate that Forrester's world model (Forrester 1971) is unable to describe the historical data of population growth. Image courtesy by Wiley

Depending on whether the "k" constant is positive or negative, "R" will either grow exponentially to infinity or go to zero with time. Of course, this is not the way that one would expect a two-species (or even a one species) population to behave.

But the Lotka-Volterra model is not "wrong," it just cannot be used in this way. Both equations of the model need to be solved together in order to make it work in a way that describes the interaction of predators and prey (although approximately). In the same way, in the "world3" model all the equations need to be solved together to make the model work as it is supposed to. Nordhaus's obvious mistake was noted and described by Forrester himself (Forrester et al. 1974):

> The case made by Nordhaus against the population sector of World Dynamics rests on the use of real-world data that he attempts to relate to model assumptions. However, Nordhaus incorrectly compares a single dimensional relationship in world dynamics (between net birth rate and material standard of living) with time series data. He fails to account for the presence of other variables influencing the time series. As a result, he erroneously asserts that the model is inconsistent with the data. In fact, the data Nordhaus present support the validity of the World Dynamics model assumptions.

Subsequently, Forrester runs his complete model and produces the following figure (Fig. 7.2):

In the Fig. 7.2, we see that the behavior of birth rates as a function of GNP produced by Forrester's model is qualitatively consistent with the historical data. Later on, Myrtveit (2005) reexamined the question and arrived at the same conclusion.

It appears clear from this discussion that Nordhaus, in his criticism of Forrester's book, had missed some basic points of the methods and the aims of world modeling by system dynamics. Unfortunately, however, Nordhaus's 1973 paper left a strong

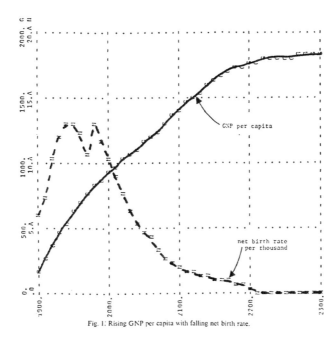

Fig. 1: Rising GNP per capita with falling net birth rate.

Fig. 7.2 Jay Forrester's response (1974) to Nordhaus (1973) criticism. This figure shows that the world model used by Forrester in his 1971 study "World Dynamics" can describe real world data. Permission granted by Springer

imprint in the successive debate, owing in part to Nordhaus' reputation, in part to the fact that Forrester's response (Forrester et al. 1974) was not so widely known and also mainly because it was published in a scarcely known journal (Policy Sciences) which was not even dedicated to economics.

On this issue, it is surprising that the editors of the "Economic Journal," who published Nordhaus's paper, did not ask Forrester to reply; as it is common policy, and even courtesy, in cases such as this one. We have no record that Forrester had asked the "Economics Journal" to publish his rebuttal, but that was the obvious first choice for him if he wanted to reply to Nordhaus, as he did. Consequently, it seems probable that the editors of the "Economic Journal" refused to publish Forrester's reply and that for this reason he was forced to publish it in another journal. This is another indication that the debate about world modeling was especially harsh and that it did not follow the accepted rules for this kind of exchange.

The scientific debate on LTG flared in the early 1970s but it gradually faltered as years went by. In 1978, Francis Sandback published a paper titled, "The rise and the fall of the Limits to Growth debate Sandback 1978" indicating that, by that time, the debate was already at an ebb. Most economists had followed Nordhaus and had decided that system dynamics was incompatible with the economics models of the time, that the LTG model had failed, and that it deserved no further discussion.

The debate about world modeling by system dynamics flared again, briefly, in 1992, when three of the authors of the first LTG book (the two Meadows and Jorgen Randers) published a sequel with the title "Beyond the Limits" (Meadows et al. 1992). In this second book, the authors updated the calculations of the first LTG study, obtaining similar results. The publication of "Beyond the limits" generated a new response from William Nordhaus; this time with the title of "Lethal Models" (Nordhaus 1992a). This new paper took up again some of the earlier arguments put forward by Nordhaus in his 1973 paper, but with considerable differences.

Facing the 43 pages of Nordhaus' 1992 paper, we immediately see that it does not contain any more the *ad personam* attacks of his first paper on this subject (Nordhaus 1973). On the contrary, Nordhaus explicitly thanks the authors of LTG for their comments and their assistance. We also see that this paper does not contain any more the accusation of "measurements without data" that was the main theme of Nordhaus's 1973 paper. All that Nordhaus has to say in this respect is (p. 14):

> In Limits I, no attempt was made to estimate the behavioral equations econometrically, although some attempt seems to have been made to calibrate some of the equations, such as the population equation, to available data.

It appears that this is not the only point where Nordhaus is backtracking. On page 15, for instance, we read that,

> the dynamic behavior of the enormously complicated LTG was not fully understood (or even understandable) by anyone, either authors or critics

And we may wonder whether with these "critics" Nordhaus also intends himself.

Nordhaus's 1992a paper is dedicated to demonstrating that the "world3" model lacks an important parameter that he identifies as "technological progress" and that, therefore, the whole LTG study is flawed.

For his demonstration, Nordhaus attempts to show that the world3 model can be recast in a simpler and "equivalent" formulation. The model that Nordhaus discusses for this purpose is nearly identical to one described by Joseph Stiglitz (1974a, b), even though Nordhaus does not cite Stiglitz's work. In turn, Stiglitz's model was derived from the one originally developed by Robert Solow in a famous paper published in 1957. The model is known today as the Solow-Stiglitz model and it is based on a concept well know to economists, that of the "Production Function."

The idea of the production function is that it is possible to describe an economic system in terms of an analytical function that includes such parameters as capital, labor, land, and others. For his purposes, Nordhaus uses a form written as (Nordhaus 1992a):

$$Y = HL^{\Omega} R^{\wedge} T^{\lceil} K^{\Delta}$$

Here, "Y" is the economic output of the system, "L" is labor, "K" is capital, T stands for "land," R for "mineral resources," and H for "human capital" or "technology." This H term is also called "total factor productivity" (TFP) or "Solow's residual" as it had been introduced by Solow in his 1957 paper. The exponents written as Greek letters are adjustable parameters that are used to fit the model to historical data.

Starting from this formulation, Nordhaus proceeds in evaluating the values of the parameters as well as the numeric values of the exponents for the whole world's economic system. For the "mineral resources" parameter (R), Nordhaus uses an exponentially decaying form as $R = e^{-\mu t}$ (even though the production of almost all mineral resources has been increasing up to now).

At this point, Nordhaus states that the model contains at least four "lethal conditions"; that is, it leads to declining production because of: (i) resources running out, (ii) insufficient land, (iii) growing population, and (iv) pollution (even though pollution is not explicitly taken into account in the model). Indeed, it is well known that this model cannot describe the historical growth of the world's economy, were it not for the "H" factor (technology) that comes to the rescue. Since this factor grows with time, the model can keep production growing even in the presence of declining mineral resources and other constraints. In the end, Nordhaus states that (p. 16):

> Technology change must exceed one-quarter of 1 percent a year to overcome the growth drag in this simple case. Historical cases of total factor productivity, h, in developed countries have been of the order of 0.01 to 0.02, which is well in excess of the rate required to offset resource exhaustion and diminishing returns.

But how is all this related to the LTG model? Nordhaus's point is that his model is "equivalent" to that of LTG because both models are "lethal" in the sense that both predict terminal decline. That is, unless they are tweaked in such a way to include a technological factor that keeps the production curves growing with time. Since the LTG model does not contain a TFP, according to Nordhaus, it cannot describe the future of the real world.

Unfortunately, there is a problem here: the two models are *not* equivalent in the sense that Nordhaus says they are. It is true that the Solow-Stiglitz model cannot describe the growth of the economy without assuming the existence of a "technology factor" (or "Solow's residual"). But the LTG model does not need anything like that. It can describe the growth of the economy without the need of an "ad hoc" factor that cannot be measured and that could be identified as "technological progress." So, Nordhaus's point is a misinterpretation of the system dynamics model used in LTG.

What Nordhaus was really doing with his 1992 paper was to present a version of the standard neoclassical model of the economy and contrast it to the LTG model. As obvious, the two models have points in common and differ in others. Both models can describe the nearly exponential growth of the economy from the beginning of the twentieth century, all the way to the present time. Where the two models diverge is, roughly, with the first decades of the twenty-first century. The Solow-Stiglitz model states that the economy will keep growing forever; the LTG model, instead, indicates that at some point it will reach a maximum and then start declining.

Obviously, some models are better than others, but a model can be proven "wrong" only by either discovering inconsistencies in its inner structure or by the simple fact that it cannot describe reality. That is what Nordhaus had unsuccessfully attempted to show for the LTG model in his 1973 paper. In 1992, he took a different approach but, again, he could not prove that the model used in the LTG study cannot describe the historical reality.

Nevertheless, Nordhaus' 1992 paper is a valuable study that examines valid points about modeling the behavior of the world's economy. It might have led to a debate and, maybe, to a definition of the competences that would pertain, respectively, to economists and to system scientists. This did not happen; it was already too late. In 1992, there was no more interest among economists in discussing the system dynamics approach that they had already decided was irrelevant at best. Indeed, the journal that published Nordhaus' paper did not ask to the LTG authors to publish a rebuttal. Instead, it gave space at the end to a "discussion" section which was extremely critical for LTG but in which the authors of LTG were not invited to participate. Again, we see that the debate on this issue was harsher than usual and did not follow the accepted rules in science.

This paper by Nordhaus sanctioned the definitive separation of two worlds: that of economists and that of system scientists. After 1992, the debate – in a scientific sense of the term – died out, with only a few papers dealing with system dynamics world modeling object written by mainstream economists (e.g., Mikesell 1995).

Reviewing the debate on the 1972 LTG study as it was conducted up to the last flare up in 1992, the general impression is that it was incomplete; that is, it had never really faced the questions that had been posed at the beginning. In most cases, criticism was based on a hasty and partial reading of the study, while some of the best known refutations of LTG, in particular those of William Nordhaus (1973, 1992a), had been based on an incomplete understanding of what system dynamics is and what it attempts to do. Only the study by the SPRU group (Cole et al. 1973) had attempted an in-depth examination of the model at the basis of the LTG study, but it had remained an isolated case. On the whole, the debate had died out not because a conclusion had been reached, but because it had not generated meaningful scientific results and, eventually, both sides had lost interest in it.

But, today, we see evident signs that the debate is reopening and that a new interest in the LTG study is appearing. It is a phenomenon that started with Matthew Simmons' paper "Revisiting the Limits to growth: Could the Club of Rome have been right?" published in 2000, which favorably reappraised the study. In the many papers which are again being published on the subject, we see the same arguments rekindled and the debate focused on two fundamental questions. Are mineral resources really finite? And how can technology help us overcome the finiteness of resources? Each of these questions will deserve a chapter in this book.

Chapter 8
Mineral Resources as Limits to Growth

Fears of "running out" of mineral resources, for instance of crude oil, periodically flare up in the debate about the economy. These fears are relatively modern as, in ancient times, the concept of mineral depletion was unknown. But, today the fact that minerals exist in limited amounts is (or should be) obvious. Then, if the minerals we use are a limited resource, there follows as a consequence that one day or another we will run out of them. And, without mineral resources, how can our civilization survive?

The question of the finiteness of mineral resources was a fundamental theme in the debate about the 1972 LTG study, which was criticized over and over on this point. One common accusation was that the input data of the LTG models were too pessimistic (see e.g., Cole et al. 1973; Nordhaus 1992a). However, this criticism could not falsify the approach of the study, which also examined scenarios in which the amounts of mineral resources were assumed to be much larger than indicated by the current estimates. Even in such assumptions, economic collapse remained a robust feature of the simulations.

But the point made by the critics of LTG was often more subtle than simply stating that resources were more abundant than assumed to be. It has to do with the very definition of "resource." A resource is, obviously, something that we can use; so, what makes a chunk of earth's crust a "mineral resource"? That depends, of course, on what we need to do, on what we can afford to do, and on what technologies we have in order to do it. Therefore it may be argued that any model which starts from the assumption of fixed and limited mineral resources is incorrect; a point that applies specifically to the world3 model used in LTG. This point was often made in the early debate (e.g., Cole et al. 1973; Simon 1981; Nordhaus 1992a) and in the modern rekindling of the discussion (see, e.g., O'Neill 2010; Radetzki 2010). This is a crucial question about the validity of the LTG model and the present chapter is dedicated to discussing it in detail.

The debate about the finiteness of mineral resources goes back to the sixteenth century. At that time, Agricola (Georg Paver 1494–1555) published his "De Re Metallica," one of the classics of geology. With that study, the ancient idea that minerals regrow as they are extracted was gradually abandoned. It resurfaces occasionally

U. Bardi, *The Limits to Growth Revisited*, SpringerBriefs in Energy: Energy Analysis, DOI 10.1007/978-1-4419-9416-5_8, © Ugo Bardi 2011

today, as with the recent diffusion of the theory of abiotic (or abiogenic) oil (e.g., Kenney 1996; Kenney et al. 2001). However, this theory has been soundly rejected by a majority of geologists (Hook et al. 2010).

Among mineral resources, fossil fuels are a crucial commodity for modern society and are crucial in obtaining nearly every other resource. The debate about the future supplies of fossil fuels, and in particular about crude oil, is perhaps the most visible and the most important today. Fears of "running out of oil" are periodically fashionable or ridiculed, depending on the vagaries of oil prices. But the debate on this issue is much older and it goes back to at least a couple of centuries.

Perhaps the first geological assessment of the finite nature of fossil fuels can be found in the book "Natural History of the Mineral Kingdom," by John Williams, published in 1789, which contained a chapter titled, "The Limited Quantity of Coal of Britain." Later on, William Stanley Jevons (1835–1882) examined the issue in his "The Coal Question" of 1856.

Jevons' essay, "The coal question," is still well worth reading today, as it echoes modern fears on oil depletion. For Jevons, it was clear that the British endowment of coal was limited but he also put the question under the light of a basic principle of economic theory: that of *diminishing returns*. The cost of extraction of coal varies depending on such factors as the quality of the coal, its depth, and the thickness of the vein. Jevons noticed that the best and easiest to extract coal was, obviously, the kind extracted first. That made coal progressively more expensive to extract. Jevons concluded that depletion would eventually make coal too expensive for the British industry. At that point, Jevons surmised that production would decline.

Jevons had not attempted to estimate when exactly the decline of the British coal industry would start, but the time frame of his analysis was of about a century or less. He says (Chap. XII):

> I draw the conclusion that I think any one would draw, we cannot long maintain our present rate of increase of consumption; that we can never advance to the higher amounts of consumption supposed. But this only means that the check to our progress must become perceptible within a century from the present time.

Indeed, Great Britain's coal production reached a maximum and started to decline less than one century after Jevons's had published his study (Bardi 2007a). Today, Great Britain produces less than one tenth of the coal it produced at the peak (Fig. 8.1).

"The coal question" was a very advanced study for its time but, under some respects, it suffered the same destiny as LTG did, more than a century later. Because of the prestige of Jevons as an economist, his views on coal depletion were never rejected or demonized, but his contribution to the field was basically ignored; unlike his contributions to marginal theory and neoclassical economics. Perhaps it was because Jevons's scenario of coal depletion spanned almost a century and people's time horizon is not so long. Or perhaps it was because the prediction that coal was going to run out, one day or another, cast Jevons in the role of just another Cassandra, ignored if not ridiculed.

Curiously, Jevons himself would be today probably one of the fiercest critics of the way that neoclassical economics has ignored the resource issues that were the

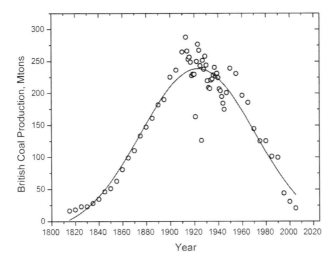

Fig. 8.1 British coal production from 1815 to 2004 (from Bardi 2007a). The production data are fitted with a Gaussian function which approximates the Hubbert curve. The maximum historical production is in 1913 with 287 Mtons, the maximum of the fitting curve is in 1923. Image courtesy of ASPO (Association for the Study of Peak oil)

basis of much of his life's work, and were even at the basis of his thoughts on marginal utility. Jevons is also the origin of the so called "Jevons paradox" which says that improvements in efficiency in the use of a nonrenewable natural resource do not necessarily lead to the resource lasting a longer time. Jevons correctly noted that a higher efficiency leads to lower prices and – therefore – to a higher consumption. The "Jevons paradox," indeed, is not a paradox at all, but it summarizes nicely one of the points made in the LTG calculations: that larger amounts of natural resources (or a better efficiency in exploiting them) do not lead to anything but a harsher collapse of the economic system.

In any case, after Jevons, the problem of mineral depletion laid dormant for decades, until it was approached again after the First World War. The war had made it clear that the supply of mineral commodities was a crucial strategic problem. The peak of British coal production, around 1920 (Bardi 2007a), can be seen as the start of the decline of the British Empire and, perhaps, one of the major factors in it. The peaking of coal production in Britain was noted but never really understood. For instance, M.W. Kirby, in "The British Coal mining Industry, 1870–1946" (1977), discusses the decline of British production but never mentions the term "depletion." In the book, Jevons is mentioned only once and in regards of matters that have nothing to do with "The Coal Question" of 1856.

Coal was not the only resource that showed depletion problems in the 1920s. At that time, doubts were expressed about the capability of the US national industry of providing enough oil for the economy. These worries turned out to be misplaced, at least at that time, but the problem could not be ignored.

It was probably because of these fears that, after the first world war, economists went back to examine the question of mineral depletion. One of the best known results of this renewed interest is the work of Harold Hotelling (1895–1973). His model, known as the "Hotelling Rule" (Hotelling 1931), was destined to have a strong impact on economic thought.

Hotelling's rule is derived from simple assumptions. About the resource, these are that: (a) it exists in a limited amounts and (b) it has zero extraction cost. Further assumptions of the model are that (c) the firm that carries out the extraction has complete monopoly of the resource, (d) that it has complete knowledge of the amount available at any moment, and (e) that it wants to maximize its revenue from extraction, called here "rent."

According to Hotelling, the lifetime of the exploitation of the resource is determined by the availability of a "backstop" resource, supposed to be more expensive, which enters the market when the price of the minerals being extracted becomes high enough to make the backstop resource competitive. At that point, the extraction of the first resource ceases.

Starting from these assumptions, Hotelling demonstrated that the firm can maintain a constant revenue over a finite time if the price of the resource rises with the current discount rate (assumed to be constant) and if production declines exponentially.

Hotelling's model is based on intuitive concepts that we can understand if we think of it as describing beer cans in a refrigerator. It is true that one would prefer a beer now rather than later but, if there is no possibility of replacing the beer cans, it is likely that one will tend to drink less as supply dwindles. In other words, beer will appear more valuable, and hence more expensive, as there is less of it.

This is the main result of Hotelling's rule and it has a certain, intuitive logic: one would expect, indeed, that the price of a resource should go up as there remains less of it. As a consequence, it is commonly assumed that prices should be a good indicator of the availability of a mineral resource. However, this assumption is often extrapolated to conclusions that the model does not support.

In many historical cases, the prices of mineral commodities have not shown the gradual growth that Hotelling had predicted. On the contrary, often a constant or downward going price trend has been observed (Barnett and Morse 1963; Slade 1982). From that, some observers arrived to the conclusion that the resource had been exploited only to a minimal fraction of the amount available (e.g., Nordhaus 1992a; Houthakker 2002). Others concluded that the resource wasn't even limited at all. Exemplary in this sense is Julian Simon who, in his book "The ultimate resource" (Simon 1981), arrived to the conclusion that the worldwide mineral resources are "infinite" on the basis of five price trends.

But, if the historical data about price trends rarely agree with the predictions of Hotelling's model, it is more reasonable to assume that the model is not valid rather than jumping to wide-ranging conclusions about an apparent great abundance of the resource. Furthermore, not only historical price trends haven't been – normally – in agreement with Hotelling's model, but also production trends haven't followed the gradual decline that the model predicts. Rather, practically all mineral commodities have shown robust growth trends up to recent times.

In practice, it may well be that stable or decreasing prices of mineral commodities do not reflect at all a great abundance of the available resources, but simply the short time horizon of the firms engaged in mining them. For the managers of a firm, worried about next quarter's report, resources can be considered as "infinite" if there is no evidence that they are going to run out in a few years. As a consequence, it makes no sense for them to optimize economic yields taking into account that, at some time in a remote future, a hypothetical backstop resource will take over. In addition, firms have no monopoly in extraction. In a free market, single firms cannot raise prices by withholding production as that would simply allow other firms to increase their market share.

So, Hotelling's rule is applicable only in the abstract world defined by its assumptions. In the real world, constant or diminishing prices of a resource are not necessarily an indication of abundance (Reynolds 1999). The rule may take hold only when the time frame of exhaustion of a resource enters the time horizon of firm managers. Only at that moment, prices will start going up because managers realize that they are going to have troubles in keeping production running. This may well be what we are witnessing in the present historical phase of increasing prices of all mineral commodities. If what we are seeing now is the result of depletion starting to bite on the budgets of the extracting firms, then we can only conclude that a misinterpretation of Hotelling's rule has lulled us in believing that mineral resources are more abundant than they really are.

But, if Hotelling's model may be seen as excessively optimistic, other models are even more so. One such model is called the "functional model," also known as the "resource pyramid." The model may have several fathers, but it was perhaps described for the first time by Erich Zimmermann in a book titled "World Resources and Industry" (1933). The main features of the model have been described in more recent times, for instance, by Julian Simon (1981) and William Nordhaus (1992a).

The functional model takes into account a fact that Hotelling had not considered: that mineral resources are not at all similar to beer cans in a refrigerator. Mineral resources are always "graded" in the sense that some are more expensive than others. The cost of a resource depends on geological factors such as depth, concentration, purity, and others.

In general, those parts of the earth's crust where a mineral is more concentrated than in the average are termed "deposits." Only some deposits are concentrated enough that we can profitably extract minerals out of them; these are called "ores." The petroleum industry uses the term "wells" for exploitable oil and gas while the coal industry speaks in terms of "beds" or "seams" for extractable coal. In all cases, anyway, terms such as "reserves" and "resources" are used to indicate deposits which can be economically exploited. Typically, the term "reserves" indicates concentrations of minerals which are known to exist and to be economically exploitable at the time of the estimate. "Resources," instead, may be used as a generic term but it may also indicate concentrations of minerals that may be exploitable at some future time or that are estimated to exist although haven't been identified yet.

Exploitable mineral resources come in different forms. Ores normally differ in terms of the concentration of the useful mineral they contain and this variation is expressed in terms of "ore grade." As obvious, higher grade ores are more profitable for mining than lower grade ones. Oil and gas wells are not "graded" in the same way as ores, and the same is true for coal seams. However, even wells and seams come in different profitability grades depending on such parameters as depth of extraction, geographical location, quality of the resource, and size of the deposit.

The functional model considers these factors. It notes that extraction starts from the most profitable resources: high grade ores. As high grade ores are exhausted, extraction progressively moves towards lower grade ores. In order to extract the same amount of mineral from a lower grade ore, it is necessary to process a larger amount of material. As Jevons had already noted in his "The Coal Question" (Jevons 1866), it is this physical phenomenon that, in the long run, pushes up the prices of minerals; not the abstract reasoning that is at the basis of Hotelling's rule.

It is at this point that Jevons and the functional model diverge. Whereas Jevons had surmised that increasing extraction costs would reduce production, the functional model assumes that, in a free market, higher prices would stimulate the extraction of low-grade resources. As a consequence, production can continue increasing.

Following this line of reasoning, the functional model emphasizes the fact that there is no such thing as "running out" of anything. This conclusion derives from "Lasky's law" (Lasky 1950) which says that the absolute amount of minerals in the earth's crust grows with dwindling ore grade. That is, the total amount of ores grows geometrically with arithmetically dwindling ore grade.

As an example consider copper. The average amount of copper in the upper crust is reported to be around 25 ppm (Wikipedia 2010). Considering a land area of 150 million km^2 and an average rock density of 2.6 g/cc, we can calculate that ten trillion tons of copper should be available within one km depth from the surface. In comparison, the USGS estimates that copper global land-based resources could be no more than three billion tons. Hence, the functional model proposes the counterintuitive effect that resources become more abundant as they are extracted. That is, we may be "running into"[1] resources rather than "running out" of them.

It is this concept that gives to the functional model the alternative name of "resource pyramid"; a graphical representation of the concept that one moves down from small amounts of high grade ores (the tip of the pyramid) towards larger amounts of lower grade resources (the lower layers of the pyramid). This phenomenon is called also "resource creation." This term is very often used in the extractive industry and it has not to be taken, of course, as meaning that mineral resources are created out of thin air. It means that new market conditions coupled with new technologies transform into exploitable resources deposits that were not economically exploitable before. This is an assumption that takes the functional model one step forward in optimism with respect to Hotelling's model which, instead, assumed gradually rising prices and dwindling production.

The functional model has a good point in considering the effect of prices on mineral extraction. Prices cause capital to be moved from one sector of the economy to another and that may offset the effect of depletion. With more equipment, workers,

---

[1]The expression "running into oil" is commonly associated with Prof. Peter Odell of Rotterdam University.

Fig. 8.2 Historical oil prices, corrected for inflation. Data from www.bp.com

and energy, it is possible to keep extracting minerals from ores of progressively dwindling grade. In this sense, the functional model can be seen as a qualitative dynamic model that considers only one element of the economy: the transfer of capital resources to the mining sector from the rest of the economy. It is implicit in the model that the transfer would take place in an ever-growing, robust economy so that the increasing needs of capital for the extractive sector can be accommodated without damaging the economy as a whole.

However, the functional model fails to address one basic question: what happens if the need of capital for the extractive sector grows more rapidly than the economy? In that case, the transfer wouldn't be painless; to say the least. To avoid this pain, the only hope is technology that is expected to keep in check extraction costs and bring prices down. Unfortunately, technology has monetary and energy costs and there are limits to what it can do; for sure it cannot extract minerals which are not there. Technology itself may be subjected to diminishing returns, as it seems to be evidenced by the fact that the prices of most mineral commodities have shown a tendency to increase during the past few decades. Oil prices, for instance, show this kind of behavior (Fig. 8.2).

The curve of oil prices can be described as "U-shaped." This behavior is by no means limited to crude oil, but it is observed for a variety of minerals (Slade 1982). It can be interpreted as the result of the competition between depletion, on one side, and of technology and economies of scale on the other. At the beginning of the extraction cycle, advances in technology and economies of scale lead to a decline of prices. There follows a phase of nearly constant prices where technology keeps improving and economies of scale increasing, but the result can only offset the higher costs of gradual depletion. At some point, prices start going up as the increasing difficulty of processing ever more dispersed resources cannot be offset any more by technological improvements or economies of scale.

Prices, eventually, are just tags stuck onto physical entities. On a barrel of oil, we may write a tag that reads $10, or $100, or even $1,000. That does not change the physical fact that the barrel can produce about 6 GJ of energy, which is what we need it for. The tag, also, does not change the fact that it takes energy in order to explore, extract, and process oil in order to transform it into that barrel. Oil extraction provides a monetary profit but, more than that, it produces an energy profit – a net gain for the economy. In the case of energy producing minerals, this gain is often measured in terms of "energy return for energy invested" "EROI" or "EROEI" (Hall et al. 2008). However, the concept that energy is needed for extraction applies to all mineral resources: the more energy needed, the more expensive – and the less profitable – extraction will be.

The role of energy in the extraction of minerals was examined in depth perhaps for the first time in a well-known paper by Goeller and Weimberg (1978) titled "The Age of Substitutability." The main points of the study can be summarized as follows.

1. Mineral resources, once exhausted, can be replaced with equivalent resources, e.g., aluminum instead of copper, synthetic oils instead of mercury, etc. provided that sufficient energy is available for extraction. The sole exception is phosphorus, which is essential for agricultural production and for which there is no possible substitute.
2. If we have sufficient energy, we can always mine lower concentration ores; at the very limit we can mine the undifferentiated crust forever.

The authors of "The Age of Substitutability" were highly optimistic about the future availability of energy; is something hardly surprising since Weimberg was head of the Oak Ridge National laboratory and an incessant advocate of nuclear power which, he believed, would lead to the availability of cheap and abundant energy in the future. On this basis, Goeller and Weimberg went on describe what they called "the principle of infinite substitutability" as follows:

> We now state the principle of "infinite" substitutability: With three notable exceptions – phosphorous, a few trace elements of agriculture and energy-producing fossil fuels (Ch_x) – society can subsist on inexhaustible or near inexhaustible minerals with relatively little loss of living standards.

This "universal" principle seems to be not so universal, since the authors themselves state that not everything can be substituted. But the important point is that the principle of infinite substitutability can work only if the economic system can produce enough energy to make it work. The question, then, is: can we realistically produce that much energy?

Energy is a physical parameter and cannot be created out of thin air, as it is possible to do with money. So, how does energy availability and cost affect the availability of mineral resources? This is a point that deserves to be examined in detail.

The earth's crust contains enormous amounts of dispersed minerals. It is not inconceivable to start from a chunk of undifferentiated crust, process it in such a way to separate all the elements in it, and use them as commodities for the economy. It could be done, for instance, using a mass spectrometer; that is ionizing the atoms and separating them for their different atomic weight. In this way, we could build a "universal mining machine" (Bardi 2008c) which could take any chunk of crust as input and produce all the elements of the periodic table, each one neatly stacked in its box.[2]

However, the boxes would contain very different amounts of materials. The Earth's crust is said to contain 88 elements in concentrations that span at least seven orders of magnitude. Some elements are defined as "common," with concentrations over 0.1% in weight. Of these, five are technologically important in metallic form: iron, aluminum, magnesium, silicon, and titanium. All the other metals exist in lower concentrations, sometimes much lower. Most metals of technological importance are defined as "rare" and exist mostly as traces in ordinary rock; that is, dispersed at the atomic level in silicates and other oxides. The average crustal abundance of rare elements, such as copper, zinc, lead, and others, is below 0.01% (100 ppm). Some, such as gold, platinum, and rhodium exist in the crust as a few parts per billion or even less.

Given these differences, a hypothetical "universal mining machine" would have to process truly huge amounts of earth's crust in order to obtain amounts of rare minerals comparable to those that we extract today from ores. The results, most likely, would not be pretty. But, the main problem would be a different one: energy costs.

The energy cost of extraction can be taken as inversely proportional to concentration. That is, extracting minerals from an ore that contains half of the concentration requires twice as much energy (Skinner 1979). Let's make a practical example. Today, we extract copper from ores, mainly chalcopyrite, $CuFeS_2$, that contain it in concentrations of around 1%. These values correspond to a concentration of copper of around 0.5% in weight. With time, the mining industry has been exploiting progressively lower grade ores but, in practice, it appears that we hit some kind of concentration barrier; as shown in Fig. 8.3 (from Meadows et al. 2004).

It may very well be that the limiting factor that is causing this behavior is the energy involved. At present, the energy needed to produce copper metal from its ores is in the range 30–65 megajoules (MJ) per kilogram (Norgate and Rankin 2002) with an average value of 50 MJ reported by Ayres (2007). Using the latter value, we find that we need about 0.75 exajoules (EJ) for the world's copper production (15 million tons per year). This is about 0.2% of the world total yearly production of primary energy (400–450 EJ) (Lightfoot 2007).

The following table lists the specific energy needed for the production of some common metals, together with the total energy requirement for the present world production (Table 8.1).

Taken together, these data indicate that the energy used by the mining industry may be of the order of 10% of the total primary energy produced. This estimation is consistent with the results of Rabago et al. (2001) who report a 4–7% range and with those of Goeller and Weinberg (1978) of 8.5% for the metal industry in the United States only.

---

[2] An extreme version of this idea was presented in a 1940 science fiction story by Willard Hawkins, titled "The Dwindling Sphere" where it was assumed that the earth's crust could be directly transformed into food.

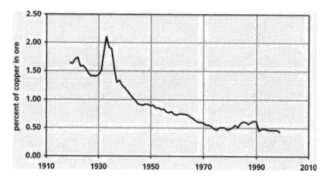

Fig. 8.3 Historical data for the ore grade mined in copper production. From (Meadows et al. 2004), Image courtesy by Dennis Meadows

Table 8.1 Specific and total energy of production for some metals. The data on the specific energy are from Norgate and Rankin (2002). Those on production are from the United States Geological Survey (USGS) for 2005

Metal	Specific production energy (MJ/kg)	World production (Mtons/year)	Total energy used (EJ)
Steel	22	1,100	24
Aluminum	211	33	6.9
Copper	48	15	0.72
Zinc	42	10	0.42
Nickel	160	1.4	0.22
Lead	26	3	0.08

It is clear that mining is energy intensive. If we were to face a reduction in the availability of energy, it would be immediately reflected in a lower availability of all minerals. At the same time, if we were forced to move to exploit lower grade ores that would immediately generate a significantly increased demand of energy that the present world's infrastructure would have a hard time to provide.

A further problem is related to the concept of "mineralogical barrier" developed by Brian Skinner (1976, 1979). That is, the distribution of most chemical elements of technological interest in the earth's crust may be "bimodal," with a very large peak at low concentration and a much smaller one at high concentration. We can mine from the high concentration peak but, as we move towards lower concentrations, we face a "dead zone" – the mineralogical barrier – where there is little or no mineral resource at concentrations useful for extraction.

If we carry this reasoning to its ultimate consequences, we can calculate the energy needed to make a "universal mining machine" running. For copper, for instance, we can estimate about 400 GJ/kg as the energy of extraction of crustal copper at a concentration of ca. 50 ppm (Bardi 2008c). If we wanted to keep producing in this way 15 million tons of copper per year, as we do nowadays, we would have to spend 20 times the current worldwide production of primary energy.

So, we can manage to extract and process large amounts of minerals only because we are exploiting energy provided for free by geochemical processes of the remote past which have created the mineral deposits we observe today and the ores we exploit. High-grade ores embed a lot of energy, either generated by the heat of Earth's core or by solar energy in combination with biological processes (De Wit 2005). But, the processes that created ores are rare and slow. From the human viewpoint, ores are a finite resource.

There is little hope of finding high grade sources of minerals other than those we know already. The planet's crust has been thoroughly explored and digging deeper is not likely to help, since ores form mainly because of hydrothermal processes that operate near the surface. The oceanic floor is geologically too recent for containing ores; only the sea floor near the continents could be a useful source of minerals (Roma 2003). The oceans themselves contain metal ions, but extracting rare metals from seawater is out of question because of their minute concentrations that makes the process highly expensive in energy terms (Bardi 2010). In addition, the amounts dissolved are not very large. For instance, considering the concentration of copper in the oceans (Sadiq 1992) we can calculate that the total amount dissolved corresponds to 10 years of the present mine production. Some suggest outer space as a source of minerals but the energy cost of leaving earth is a major barrier. Then, most bodies of the solar systems – e.g., the Moon and the asteroids – are geochemically "dead" and contain no ores.

Recovering minerals from waste is not a solution, either, at least in the present situation. Normally, industrial and urban waste is either dumped in landfills or incinerated. In both cases, the result is that postrecovery is nearly impossible (Shen and Fossberg 2003). About recycling, according to the USGS (Papp 2005), in the United States the average recycling rate is about 50% in weight for the principal metals produced. This is not enough for compensating the decline of ore grade in mining.

So, if we are to solve the problem of ore depletion, we need a radical change in the way industrial production is conceived and managed. That, in turn, would require investments that are unlikely to materialize until the shortage of minerals becomes very serious. But, at that moment, the resulting decline of the economy would prevent us to muster sufficient capital to build up a completely new infrastructure.

We can now go back to the original question that had started this chapter and examine how mineral depletion was modeled in LTG. We see that the basic assumptions of the world3 model are compatible with the considerations developed here. We read in the 2004 edition of LTG (p. 145)

> As resources are consumed, the quality of the remaining reserves is assumed to decline <..>
> That means more capital and energy will be necessary to extract, refine, and transport a ton
> of copper or a barrel of oil from the earth. Over the short term these trends may be offset by
> technological advance. Over the long term they will reduce the capacity for physical
> growth.

If this behavior is examined in a simple dynamical model that considers only resources and capital (Bardi and Lavacchi 2009), the result is that the production of the resource follows a "bell shaped" curve similar to the one proposed by Hubbert

(1956) for crude oil on the basis of an empirical approach. The behavior of several mineral commodities also shows the "bell shaped" production curve (Bardi and Pagani 2007), indicating that these models correctly describe the extractive process in a market economy.

The world3 model of LTG does not examine single mineral resources but aggregates all of them into a single "non-renewable resources" parameter. It is assumed that these resources exist in finite amount and that the rate of extraction depends on the available capital of the extractive industry. The economy can divert some capital from other sectors to the extractive industry, but this capital is not infinite. This approach is clearly consistent with the considerations developed here and with the principles of physics. The results for the base case scenario (and for many other LTG scenarios based on reasonable physical principles) – the natural resources parameter declines with time and production shows a peak.

At this point, we can answer the question that was asked at the beginning of this chapter: is it correct to assume that minerals are limited resources? The answer is yes, provided that we understand that the limits we are facing are not limits of *amounts*, but limits of *energy*. Most of the energy we use comes from mineral resources (fossil fuels and uranium), which are limited in amounts and may be already in decline. The rest comes from renewable resources which have an upper limit to production determined by the energy that comes from the Sun.

Of course, an unexpected technological breakthrough in energy production could change the picture, but we cannot bank on that happening soon, or even at all. Hence, the depletion of our energy resources combines with the depletion of mineral ores to ensure that in the future we will see a decline in the production of mineral commodities – a decline that may have already commenced for a number of minerals (Bardi and Pagani 2007). In the end, the fact that we live in a limited world was the basic assumption at the core of the LTG scenarios and the considerations developed here strengthen the soundness of that assumption.

Chapter 9
Technological Progress and Limits to Growth

In 1971, President Nixon signed the "the National Cancer Act" declaring what came to be known as "The war on cancer." In the wake of the enthusiasm brought by the success of the Apollo program, which had taken men to the Moon and back in 1969, it seemed possible to repeat the same approach in the fight against cancer; solving the problem by concentrating scientific, technological, and financial resources on it.

Today, after 40 years and more than a 100 billion dollars spent in the effort (Kolata 2009), much progress has been made in the cure of cancer (Mukherje 2010), but the results have been far less spectacular than those of the Apollo program (Spector 2011). Cancer remains a major cause of death in the Western World.

The story of the effort against cancer shows that technological progress is not always the magic bullet that cures all ills and solves all problems. Yet, our times are much different from those of Edward Gibbon who, in his "Decline and fall of the Roman Empire" (1788), was perhaps the first to note that, "The ancients were destitute of many of the conveniences of life which have been invented or improved by the progress of industry." For us today, progress is a normal feature of life; it may not bring the miracle cure for cancer, but we see it happening and it constantly changes the way we do things. We expect that progress will improve our lives.

One of the most common forms of criticism against "The Limits to Growth" study was that it did not take into account technological progress. But, as we saw in the previous chapters, this is not true. The LTG models did develop scenarios that implied assumptions that could be only the result of technological breakthroughs; for instance their use of infinite or very large amounts of natural resources. But one could still say that LTG did not take into account technological progress in the right way. And that brings us to a basic question: what is, exactly, progress? Clarifying this point is essential for the debate about the validity of the LTG approach and it is the subject of the present chapter.

We all know that progress means that things become better. Planes go faster, TV screens become bigger, cell phones become smaller, and so on. But, how to transform these observations into parameters that we can input into the equations of a mathematical model? What is that we should measure? Speed, size, complexity, or what?

U. Bardi, *The Limits to Growth Revisited*, SpringerBriefs in Energy: Energy Analysis,
DOI 10.1007/978-1-4419-9416-5_9, © Ugo Bardi 2011

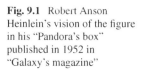

Fig. 9.1 Robert Anson
Heinlein's vision of the figure
in his "Pandora's box"
published in 1952 in
"Galaxy's magazine"

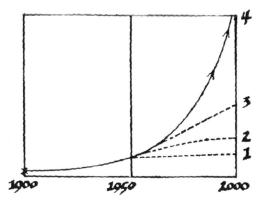

These difficulties have not deterred people from seeking for some methods
that could be used for modeling progress. Robert Anson Heinlein, well known as science
fiction writer, was perhaps the first who attempted this task with his 1952 article
titled "Pandora's box" (Heinlein 1952). In that article, we find the figure above
(Fig. 9.1).

Of the four possible cases shown, Heinlein was enthusiastically favorable to case
4 – exponential growth of progress – even though he couldn't explain why, nor what
was being plotted, exactly, in this evidently hand-made figure.

Heinlein accompanied his interpretation of progress with a number of detailed
predictions for what the world should have looked like in the year 2000 that is
50 years afterwards. This kind of predictions, called "technological forecasting,"
are notoriously very difficult when projected over long time spans and Heinlein's
predictions turned out to be nearly all wrong. He had foreseen the development of
antigravity, space flight for the masses, life extension over 100 years for humans,
and many other wonders that never materialized. On the contrary, he failed to imagine
such things as the Internet, cell phones, personal computers, and most of what we
consider today as the tangible manifestation of progress.

Heinlein's mistakes highlight how difficult it is to predict the future. Nevertheless,
his exponentially growing progress curve was a brilliant intuition. His model could
not be quantified but, at least, he had focused on the right question and he had given
an answer that was to prove popular in the following years. For instance, in the book
by Damien Broderick "The Spike" (Broderick 2001), we find again the same concept
of exponential growth of progress and the same problem of quantifying it.

The question of progress was examined in a more quantitative way a few years
after Heinlein's "Pandora" paper, when the economist Robert Solow published the
results of a study that is often considered the basis for the understanding of technological progress in economics (Solow 1957). It is likely that Solow did not know of
Heinlein's work and surely he had a very different approach. Yet, Solow arrived to
the same conclusion: progress grows exponentially with time.

Here is an illustration from Solow's original 1957 paper, where we see the plot of
a function called "A(t)" which Solow suggested was to be understood as a quantitative

Fig. 9.2 Robert Solow's famous 1957 graph showing the "A(t)" function, known also as "Solow Residual" that he proposed should be identified as a quantitative measurement of the productivity improvement generated by technological progress

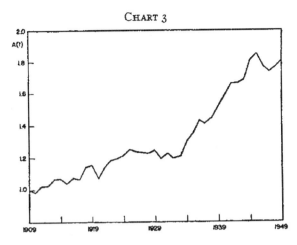

measurement of technological progress. In the figure, the A(t) function is expressed as a fraction of the value for 1909, taken as unity. It grows exponentially with time at a rate of 1–2% per year (Fig. 9.2).

Solow arrived at this concept while trying to describe how the gross domestic product (GDP) of the United States varied with time. He was using a "production function" in the form called "Cobb-Douglas" from the names of the authors who first proposed it in 1938. We have already seen in a previous chapter how Nordhaus (1992a) used the concept in order to propose an alternative model to the LTG one. Here, it is worth to go more in detail in this theory.

The production function can be written in its simplest form as (Ayres 1998)

$$Y = L^a K^b,$$

where Y is production, L is labor, and K is capital, while "a" and "b" are exponents sometimes called "elasticities." This simplified form already shows some of the fundamental characteristics of this function. For instance, it is normally (but not always) assumed that the factors are to be "homogeneous," a condition equivalent to satisfying Euler's theorem. In this case, the system displays constant returns to scale; i.e., doubling both factors (L and K) doubles the output. This condition is satisfied if $a + b = 1$. Note that the same production can be obtained with different sets of values for capital and labor. The set of values for capital and labor that produce the same output is defined as an "isoquant."

As shown here, the function is very simple, but it can be modified in order to take into account more factors, as William Nordhaus (1992a) did with the aim of refuting the LTG approach. In the 1950s, however, Robert Solow was using the function in its simplest form as a way to describe the output of the US economy. He found (Solow 1957) that the function could not describe the historical data unless he multiplied it by a factor "A(t)" that later came to be known as "Solow's residual" or "Total Factor Productivity" and, in general, a quantitative measurement of how technology improvements lead to gains in productivity.

On the basis of Solow's idea, it has been often suggested (see, e.g., Nordhaus (1992a) and (Marxsen 2003)) that the exponentially growing progress curve will continue to grow at the same rate in the future and, therefore, compensate for resource depletion as it has done in the past. Indeed, Solow himself is reported to have said that, "The economy, in fact, can carry on even without natural resources" (Solow 1974) although that was to be seen, of course, as an extreme possibility.

This view of progress changes at its basis the structure of the world models of LTG. If an exponentially growing "technology" factor is added to the model, and if this factor is sufficiently large, then it is possible to obtain scenarios where the growth of industrial and agricultural production keeps going on forever or, at least, as long as the progress parameter is assumed to maintain the exponential trend. This possibility was shown independently by Boyd (1972), Cole and Curnow (Cole et al. 1973) and by the authors of the LTG study themselves with the model they called "IFI-IFO" (Meadows et al. 2004). The need of such a factor is also the gist of Nordhaus's criticism of LTG (Nordhaus 1992a).

Models that compensate resource depletion with human ingenuity may be comforting, but how reliable are they? One problem is that the "Solow residual" is by no means just a residual. It is a major component (half or more) of the production function which would not be able to describe the historical results without it, not even qualitatively (Ayres 2001, 2006). And that is not the only problem: Solow may have considerably overestimated the role of progress. Hall et al. (2001) showed that, if the growth in energy production is taken into account, the "residual" basically disappears. Herman Daly strongly criticized (Daly 1997) Solow approach. We also read in Daly's "Steady state Economics" (Daly 1977) (chapter 5):

> The idea that technology accounts for half or more of the observed increase in output in recent times is a finding about which econometricians themselves disagree. For example, D. W. Jorgenson and Z. Grilliches found that "if real product and real factor input are accurately accounted for, the observed growth in total factor productivity is negligible" (1967). In other words, the increment in real output from 1945 to 1965 is almost totally explained (96.7 percent) by increments in real inputs, with very little residual (3.3 percent) left to impute to technical change. Such findings cast doubt on the notion that technology, unaided by increased resource flows, can give us enormous increases in output. In fact, the law of conservation of matter and energy by itself should make us skeptical of the claim that real output can increase continuously with no increase in real inputs.

A further perplexity on the role of Solow's residual derives from the fact that it may be the only entity in economics that is supposed to keep growing forever. That is curious, to say the least, considering the established concept of "diminishing returns" in economic sciences. Why should technological progress be exempt from this very general law? This point was examined, for instance, by Giarini and Laubergé (1978) specifically in view of discussing the basic assumption of the LTG model. Their conclusion was that technology does show diminishing returns. That is, increasing investments in research does not bring proportional results in terms of technological improvements.

This conclusion makes sense if we examine practical cases of progress. There are many technologies which have literally exploded in spectacular phases of progress but have slowed down afterwards. Think of how automobiles quickly grew from clumsy self-propelled carts in mid nineteenth century to models that were already modern in their basic features at the beginning of the twentieth century, such as the Ford model T which entered production in 1908. After this period of rapid technological growth, money continued to be spent in R&D for the automotive industry and it is still spent today. But this money is not creating any more the kind of revolution in transportation that led to the Ford T. In the 1950s, some people expected flying cars to become commonplace, but that did not happen and probably never will. In short, R&D on cars has followed a clear path of diminishing returns.

We see the same path not just for cars. There are plenty of technological areas that are progressing very slowly, if they are progressing at all and not for lack of investments. Just think of how the human average life expectancy is not significantly increasing any more after the spectacular rise observed up to not long ago.

Of course, today there are technological sectors where R&D is paying back handsomely the money spent and where progress is fast occurring. We can think of communications, data processing, robotics, and others. "Moore's law" is well known: it says that the number of elements placed on a computing chip is doubling every 2 years, approximately. A consequence is that computing power is increasing at an approximately exponential rate. But we are rapidly approaching physical limits to the size of the transistors which, in the end, cannot be made smaller than single atoms. So, how long can the trend continue?

There is evidence that technological progress, considered as a whole, may be slowing down. Robert Ayres has published several papers on the subject (e.g., in 1998 and 2001) utilizing some assumptions on the general efficiency of technological processes and arriving to the conclusion that the increase of the "technological factor" (or Solow's Residual) is not exponentially growing but rather it follows a logistic (i.e., asymptotic) curve. In other words, progress, just as most of the entities defined in economics, may show diminishing returns.

Ayres's work has been confirmed by a number of authors who have reported a stalling or a diminution in the number of patents published worldwide or in the ratio of expenses per patent (Tainter 1996; Huebner 2005; Hogan 2006; Nordmann 2008; Staniford 2010). The dwindling number of patents is not, in itself, proof that technological progress, as a whole, is slowing down. But it is at least suggesting that something is wrong with the hypothesis of an ever growing exponential progress of science and technology.

This is, obviously, a very complex issue that cannot be said to be really understood. However, it may be a good idea to bring technological progress in line with other known factors of the economics theory. That is, we should assume that progress as a whole is affected by diminishing returns. So, we cannot bank on the "total factor productivity" – or "Solow's residual" – as a continuous function that will grow forever, and therefore will be able to compensate for resource depletion, pollution, overpopulation, and all the problems that are affecting us nowadays.

But there exists another view of technological progress that is much grander and more ambitious than the one that derives from the smooth curves of economics models. As an example of this view, we can cite Kevin Kelly's book "Out of Control"

(Kelly 1994) where we find a description of progress that was produced as a direct criticism of the LTG book. We read at p. 575 that:

> Direct feedback models such as Limits to Growth can achieve stabilization, one attribute of living systems, but they cannot learn, grow or diversify – three essential complexities for a model of changing culture or life. Without these abilities, a world model will fall far behind the moving reality. A learning-less model can be used to anticipate the near future where coevolutionary change is minimal; but to predict an evolutionary system – if it can ever be predicted in pockets – will require the exquisite complexity of a simulated artificial evolutionary model.

And:

> The Limits of Growth cannot mimic the emergence of the industrial revolution from the agrarian age. "Nor," admits Meadows, "can it take the world from the industrial revolution to whatever follows next beyond that."

Kelly believes that progress occurs by leaps and bounds; true quantum leaps that bring human society up to a ladder of increasingly complex stages. A further example of this attitude is expressed in the following paragraph (Ordway 1953) reporting the words of the president of an oil company:

> The fact seems to be that the first [resource] storehouse in which man found himself was only one of a series. As he used up what was piled in that first room, he found he could fashion a key to open a door into a much larger room. And as he used up the contents of this larger room, he discovered there was another room beyond, larger still. The room in which we stand at the middle of the twentieth century is so vast that its walls are beyond sight. Yet it is probably still quite near the beginning of the whole series of storehouses. It is not inconceivable that the entire globe-earth, ocean and air-represents raw material for mankind to utilize with more and more ingenuity and skill.

Such "quantum leaps" or "new rooms" may not be incompatible with a smooth, exponentially growing progress curve. At some point, it may be argued, Solow's exponentially growing function will be rising so fast that – from the human viewpoint – it will appear to, literally, shoot out to infinity. This phenomenon is sometimes described as the "Singularity" (Kurzveil 2005) or the "Omega Point" or "The Spike" (Broderick 2001). In some interpretations, this phenomenon will lead humankind to transcend into a nearly godlike, "transhuman" status.

This idea may be very old, but it seems to have been expressed for the first time in its modern form with Robert Ettinger's book "Man into Superman" (Ettinger 1974). Later on, the concept of transhumanism was popularized by Vernor Vinge (1993) and by books such as "The Great Mambo Chicken" by Ed Regis (1994) and "Cyberevolution and Future Minds" by Paul and Cox (1996). The most recent proposer of the concept of technological singularity is probably Ray Kurzveil (2005).

We cannot dismiss the possibility of humans becoming superintelligent, "transhuman" creatures. But even such creatures have to face physical limitations, at least as long as they exist on a finite planet. So, it is not enough to talk glibly and inspirationally about "quantum leaps," "co-evolutionary change," or "new rooms." If quantum leaps in technological progress are supposed to invalidate the whole approach of the LTG study – and not just postpone the unavoidable – we need to be much more specific. What quantum leaps can save us from pollution, resource depletion, and overpopulation? What kind of breakthrough do we need in order for the economy to be able to keep growing, if not forever at least for a long time?

Curiously enough, the tool for answering this question is just the modeling method used in the LTG study; the very object of criticism from technological optimists. It is with the LTG model, coupled with common sense and basic physics, that we can examine how the effects of technological breakouts may change the destiny of the human civilization.

A good example on this point comes from considering energy. What would happen if we were to stumble upon a new source of energy, both cheap and abundant?

The LTG model does not consider energy as a disaggregated parameter but we know that, with more energy, it is possible to extract minerals from less concentrated ores. Hence, the final result is a higher flux of resources into the economy. A simulation along these lines was performed already in the first LTG book (Meadows et al. 1972) and it was confirmed in the later versions. The result is that the availability of more abundant resources postpones collapse but generates it anyway as the result of a combination of overpopulation, depletion of agricultural soil, and pollution.

Of course, one may argue that part of the abundant energy available in this scenario can be used for fighting pollution – such as CO_2 causing global warming. This is, again, something that can be simulated and that has been done in the various versions of the LTG studies. One scenario assumes that pollution can be reduced by a factor of 4 with the increased resources available from a technological breakthrough in energy. Even in this case, however, collapse cannot be avoided. It can only be pushed forward for some years in the future.

Then, of course, we may also suppose that, with sufficient energy, *all* pollution can be eliminated. In this case, we are back to the IFI-IFO scenario described earlier on. But is it a physically realistic scenario? The second law of thermodynamics tells us that no system of energy production can be 100% efficient and the result is waste heat. So, if we keep increasing the amount of energy produced on earth, eventually we would face a problem of thermal pollution – heating the planet as the direct result of our activities. Pollution is a necessary consequence of everything we do – including fighting pollution.

If we want to keep pushing the technological option, we may think that we could also remedy thermal pollution by using technology: maybe giant mirrors in space would do the job; or perhaps we could use planetary thermal engineering based on controlling the concentration of greenhouse gases. But fiddling with the mechanisms that keep the earth's ecosystem running is risky, to say the least. Besides, we are piling up one assumption after another about possible future technologies that, however, do not exist at present and may never exist in the future.

In the end, we can say on this matter that new technologies may just as easily worsen the human situation rather than improving it in terms of preventing collapse, as it can be shown by playing with simulations based on the world3 model. It is true, however, that these considerations are strongly linked to the assumption that humans are confined to the surface of the earth. This is not necessarily true and moving to space would push forward the physical limits to human expansion.

Fig. 9.3 An artistic rendering of the concept of Gerard O'Neill's space habitats. Immense pressurized cylinders that could be located at one of the stable Lagrange points of the Earth–Moon system, such as the L5 one. According to Gerard O'Neill (1974) such habitats could host millions of human being and ease the problem of the overpopulation on Earth. This image was created by Rick Guidice for NASA in 1976 (http://settlement.arc.nasa.gov/70sArt/art.html)

The idea has been around for a long time and perhaps the most detailed scheme for colonizing space was proposed in 1974 by Gerard O'Neill; in part also as a response to the first edition of "The Limits to Growth." O'Neill's concept was based on immense pressurized habitats that would be placed at the stable Lagrange points of the Earth and Moon system. The points called "L4" and "L5" are naturally stable and would be the best choice for a space colony. Indeed, O'Neill's idea came to be associated to the "L5 point" (Fig. 9.3).

Some dreams of space colonization turned out to even grander. Already in 1960, Freeman Dyson (1960) had proposed to rearrange the whole Solar system by dismantling the planets and placing the obtained matter into an immense sphere around the Sun that would collect all the Sun's energy and permit humans to exploit it for their purposes. If such a feat were possible, it would increase the area of the human habitat to an enormous factor in comparison to occupying the surface of just one planet. Some other studies even considered the possibility of colonizing the whole galaxy (see a review by Tarter 2003) – a feat that could be performed in times of the order of a million years even moving at speeds lower than light.

Although LTG had not considered expansion into space, the world3 model can be modified in order to simulate this possibility. In 1976, Peter Vajk did just that, generating scenarios that took into account the much larger stocks of matter and of energy that could be available in space. He worked on O'Neill's scheme, assuming that it would be possible to build a civilization based only on materials abundant in space bodies, such as the moon and the asteroids: oxygen, silicon, iron, aluminum, magnesium, titanium, and a few others. In the reasonable assumption that pollution would be a much smaller problem in the vast region which is outer space, Vajk found that the economy could continue growing throughout most of the twenty-first century. He also found that the problems of overpopulation on earth would be solved; that would occur in part by migration to outer space and in part by the natural reduction of birth rates that comes when the quality of living improves.[1]

Maybe our destiny really is of colonizing space – first L5, then the whole galaxy. But do we have the resources needed? With the world's economy that appears to be on the verge of the collapse envisaged by the LTG "base case" scenario, it is hard to think that we can now restart space exploration and develop it fast enough to overcome mineral depletion, climate change, and overpopulation. Today, the space race has lost its momentum and it is reduced to a pale shadow of what it was in the 1960s. The lunar expeditions that had started in 1969 would be far too expensive today. With the space shuttle program winding to a close in 2011 (Atkinsons 2010), it seems clear that we cannot afford any more the kind of investments that were possible in the 1960s. And it is likely that we would require a much more massive investment if we were to build an infrastructure in space that could use the moon and the planets as new stocks of raw materials. It seems inescapable, as things stand now, that we have first to solve the problems we have here, on earth, and only later resume space travel with the idea of colonizing the solar system and, who knows, even the galaxy.

If outer space is commonly seen as the new frontier for humankind; there is at least another one: that of virtual space. As for all frontiers, in order to exploit it we must find a way to move there. That could be done using the technology of "mind uploading," or "brain emulation" (see, e.g., Kurzveil 2005; Sandberg and Bostrom 2008). The idea is to transform human minds into programs that run inside computer memories. In this way, human beings could be effectively transferred into virtual space as simulated entities. "Running" one such virtual human would require much less energy and resources than what is needed to support a real human being. That would lead to the possibility of plenty of "new room" in virtual space. Many billions of humans could live as programs in virtual space; perhaps even trillions or quadrillions.

Surely, transferring a human being into virtual space would be less expensive in terms of energy required than moving him or her to outer space. But, we are facing possibilities whose consequences are difficult to imagine. Supposing that human beings can really be transferred to virtual space; would that stop population growth in real space? Would you really be able to convince people to leave their flesh bodies in exchange for virtual ones? Besides, computers need resources to run. If we want to run trillions or quadrillions of virtual human beings as programs, we face the

---

[1] Later on, Vajk was victim of the legend of the "wrong predictions" of LTG and he backtracked from his model saying that it was to be intended as "a parody" (Vajk 2002).

same problem that we have now in keeping alive the near seven billion of real human beings: limits of the available resources.

So, we are facing a fundamental contradiction between the human desire to have more, probably ingrained in the way our brains are built (Hagens 2007) and the fact that the world's resources are limited; a consequence of the way the universe is built. The "predicament of mankind," as defined by the Club of Rome (1970) at its foundation, remains an unsolved problem. The only thing we can say with certainty is that there are no magic technologies that can lead us out of the impasse. The only way we have is to learn to live with the limits we have.

Chapter 10
The Political Debate

The publication of "The Limits to Growth" in 1972 had a very strong impact on the public opinion; so much that some feared it would spread out of control. For instance, K.R.L. Pavitt wrote about LTG: "…. neo-Malthusians are bound to increase in numbers and to be right about the end of economic growth sometime in the future. In the industrially advanced countries, this should be relatively soon" (Cole et al. 1973, p. 155).

That did not happen, of course. Instead, in the late 1980s and 1990s the world witnessed a rejection of the LTG study so complete and pervasive that it is surprising for us to discover quotes such as this one. With the 1990s, it became politically incorrect even to mention the LTG study, unless it was to state that it was wrong or had been discredited.

Only in recent times a reappraisal of the LTG study seems to be in progress, but the negative attitude is still widespread and probably prevalent. The question, therefore, is how a work that was initially hailed by some as a world-changing scientific revolution ended up being completely rejected and even ridiculed. Finding an answer requires the examination of the old literature in order to pinpoint the turning points of the battle of ideas that took place from the early 1970s to the present day.

Criticism against the first LTG book, which appeared in 1972, was rarely based on purely scientific arguments but, rather, on a "gut reaction" that was often a manifestation of the "Cassandra effect," the tendency of people to disbelieve bad news.

As an example of this emotional reaction, which arrived even from scientists and researchers, Paul Krugman reports how he and his research advisor of the time, William Nordhaus, reacted to the publication of Jay Forrester's "World Dynamics" in 1971.

> The essential story there was one of hard-science arrogance: Forrester, an eminent professor of engineering, decided to try his hand at economics, and basically said, "I'm going to do economics with equations! And run them on a computer! I'm sure those stupid economists have never thought of that!" And he didn't walk over to the east side of campus to ask whether, in fact, any economists ever had thought of that, and what they had learned. (Krugman 2008)

U. Bardi, *The Limits to Growth Revisited*, SpringerBriefs in Energy: Energy Analysis, DOI 10.1007/978-1-4419-9416-5_10, © Ugo Bardi 2011

This is a good example of the defensive reaction that was generated by LTG in several scientific fields, in particular in economics. Krugman does not address the question of whether Forrester was right or wrong – he just resents the fact that a non-economist had dared doing economics.[1]

In many cases, this reaction led to a polarized debate with a clearly detectable, although rarely explicitly stated, political background. At times, instead, the political nature of the criticism was obvious even in articles in scientific journals. Defining the LTG authors as interested in reviving "Malthusian" concepts was one such political accusation, seen, – for instance in Nordhaus (1973) and in Pavitt (Cole et al. 1973).

An example of a typical case of "split personality," political and scientific, in the reaction to LTG can be found in "Models of Doom" (Cole et al. 1973). The book is divided into two sections, one dedicated to a scientific examination of the methods and of the results of the 1972 LTG study, the other titled "The Ideological Background." Even the curators of the book seemed to suspect that the discussion in this section was going somewhat out of control. The author of the preface to the book, Asa Briggs, says (p. 9 of the US edition) that "It was suggested to us that, because of their more speculative and controversial character, these chapters ought to be omitted."

But the five essays of the "Ideological Background" section were not omitted and there the discussion veered into political accusation and outright conspiracy theories. The authors of LTG were accused of having hidden motives, of disregarding the interest of the poor, of working for a world dictatorship, of wanting to abolish democracy and substitute it with a "technocracy" and, in the final essay by Marie Jahoda (titled "Postscript on Social Change"), of "irrational millennial mood" and "escapism."

This section of the "Models of Doom" is, by now, completely obsolete and it is of little interest for us except as an indication of the attitude that was developing in the debate from the very beginning. Emotions ran high and that led many people to be carried away, losing contact with the normally accepted scientific standards of the debate.

A good example of a scientific debate gone awry is the paper by Robert Golub and Joe Townsend published in "Social Studies of Science" (Golub and Townsend 1977). According to the authors, the LTG study was the result of the attempt of multinational companies to acquire control of the world's economy, even at the cost of destroying nation states. For this purpose, they say, multinationals set up the Club of Rome as a lever to propose, and perhaps in the future impose, a series of "no growth" or "zero growth" policies which would repress all attempts on the part of the developing nations to grow out of their condition of subordination. According to Golub and Townsend, the work of the Club of Rome could eventually generate, "a kind of fanatic military dictatorship."

---

[1]Note also that Krugman seems to have forgotten that, when Forrester had the idea of "doing economics with equations" in the mid 1960s, digital computers where still such a novelty that Forrester had to use a special computer he himself had developed. Therefore, it is unlikely that any economist could have "thought of that" or "learned" about that earlier than Forrester.

It goes without saying that Golub and Townsend could not provide proof for their accusations. To give some idea of the level and the tone of the article, in it we find paragraphs such as this one:

> Limits to Growth predicted an energy-cum-pollution-and-population crisis in the next thirty to fifty years. Remarkably, a crisis seemed to appear in less than two years. In fact, the oil companies were preparing some sort of crisis long before Limits appeared.

This statement is a good example of how much reality can be bent to suit a conspiracy theory. If the LTG group had developed their models to predict a crisis that was being artificially created, they could surely have rigged their calculations in such a way to predict the crisis for when it was to appear and not for 30 or 50 years later. That such an article could be published in a refereed journal shows how the debate had gone out of control.

With time, the political discussion about LTG became harsher and more hostile with some very aggressive criticism of LTG appearing in the 1980s. Julian Simon attacked the study in his book "The Ultimate Resource" (1981) where he accused Aurelio Peccei of being "a liar" and of having invented the LTG study in order to promote his program of world domination. Two years later, Lyndon Larouche published a book titled "There are No Limits to Growth" (1983) in which – among other things – he accused the Club of Rome of being a secret society whose purpose was the extermination of the "darker races."

It is difficult today to assess exactly how effective these attacks were in demolishing the prestige of the Club of Rome and of the LTG study. In the 1980s, Julian Simon had gained a certain reputation as "doomslayer" and his books sold well with the general public. Lyndon Larouche, instead, remained always the Carrier of extreme fringe opinions. However, his 1983 book may have been the origin of a legend which, by now, can be easily found on the web and that accuses Aurelio Peccei of having commissioned the manufacturing of the AIDS virus as a way to reduce the world's population. It goes without saying that Larouche could not bring any proof for these accusations.

On the whole, in the 1980s the issues raised by the LTG study seemed to be marginalized by the optimistic mood that started to be widespread as a consequence for the end of the oil crisis that had started in the early 1970s. With the mid 1980s, oil prices collapsed with crude oil flowing to markets from the wells of the Middle East and from the North Sea. Most mineral commodities had followed the trend of crude oil in terms of prices and the bet that Julian Simon made with Paul Erlich in 1980 about future resource availability had a considerable media coverage (Regis 2004), including an especially in a derisive article that was published by Tierney in 1990 on the New York Times.

All that may have caused a shortcut in people's memory. Those who had read "The Limits to Growth" in the early 1970s vaguely remembered that the book had predicted a catastrophe that would take place at some moment in the future as the result of oil, or other minerals, running out. If the 1973 oil crisis had looked to many as the crisis predicted by LTG, the end of the crisis, in the mid 1980s, seemed to be the refutation of the same predictions. Not many people seemed to have the time or the willpower to sort out that old book from their bookshelves and check whether this interpretation was correct.

For instance, in 1987, George J. W. Goodman, (Goodman 1987) (writing under the pseudonym of "Adam Smith") managed to make at least five (!) errors in a single sentence about LTG when he wrote in "The New York Times" that "Ten years ago the Club of Rome – and a number of other economic prognosticators – predicted shortages of everything. Oil would go up to $200 a barrel and the planet would starve.[2]"

So, LTG might simply have been forgotten in the mood of general optimism of the late 1980s and 1990s, were it not for a specific event that seems to have catalyzed the aggressive demolition of the study. This "tipping point" in public perception arrived when, in 1989, Ronald Bailey published an article titled "Dr. Doom" in Forbes. The title was referred to Jay Forrester, the father of system dynamics, but the attack was, mainly, against the LTG book which Bailey described as, "as wrongheaded as it is possible to be." To prove this statement, Bailey said that:

> "Limits to Growth" predicted that at 1972 rates of growth the world would run out of gold by 1981, mercury by 1985, tin by 1987, zinc by 1990, petroleum by 1992, copper lead and natural gas by 1993.

Later on, Bailey repeated the same criticism in his book "Ecoscam" published in 1993. This statement had an incredible success and we can still see it today repeated verbatim or with minimal variations all over the web. It was the starting point of the downfall of LTG that – from then on – started to be widely described as "wrong" in the press.

Now, where had Bailey found these "predictions" that he attributes to LTG? In the text of the 1972 book, no such statements can be found. Rather, Bailey's had revisited an early misinterpretation of LTG (Passel et al. 1972) which dealt with the data of Table 4 of the second chapter of the 1972 LTG which lists the projected duration of some mineral resources calculated in some assumptions. The table (including the references) takes up five pages of the book. Here, a simplified version is reported, showing only the resources mentioned by Bailey in his 1989 article (Table 10.1).

As we can see, there are six columns in the table (further columns in the original provide details on the geographical distribution of the resources). The amount of known reserves is listed according to USGS data. Then, the projected duration of the reserves is calculated according to three assumptions:

– Column 3: production continuing at a constant rate ("static index").
– Column 5: production increasing at an exponential rate (as given in column 4).

---

[2] (1) That of confusing the "Club of Rome" with the actual authors of LTG, (2) that the book had been written 10 years before instead of 15, (3) that LTG had predicted oil prices (4) that the LTG study had predicted shortages by the time Goodman was writing and (5) that LTG had predicted famines to occur soon.

Table 10.1 Simplified Table 4 of the 1972 LTG book

1	2	3	4	5	6
Resource	Known global reserves	Static index (years)	Projected average rate of growth (% per year)	Exponential index (years)	Exponential index calculated using five times known reserves (years)
Gold	353×10^6 troy oz.	11	4.1	9	29
Mercury	3.34×10^6 flasks	13	2.6	13	41
Tin	4.3×10^6 tons	17	1.1	15	61
Zinc	123×10^6 tons	23	2.9	18	50
Petroleum	455×10^9 bbls	31	3.9	20	50
Copper	308×10^6 tons	36	4.6	21	48
Lead	91×10^6 tons	26	2	21	64
Natural gas	1.14×10^{15} cu ft.	38	4.7	22	49

– Column 6: production increasing at an exponential rate, as in column 5, but with reserves increased of a factor of five.

The original table lists many more mineral commodities, but those reported here are those that Bailey chose. He also selected only the duration data of column 5. As a result, summing up the numbers reported in the table to the date of publication of the book (1972) we find the "predictions" reported by Bailey for when these resources (gold, mercury, tin, etc..) should have run out.

But why did Bailey (and earlier on Passel and the others) choose just these commodities and just the numbers of column 5 and not those of column 3 or of column 6? Of course, because only these minerals and the numbers of column 5 permitted them to "prove" that the LTG "predictions" were excessively pessimistic or wrong.

Apart from cherry picking the data, the problem with Bailey's interpretation is that nowhere in the book was it stated that the numbers were supposed to be read as predictions. It should be obvious that Table 4 was there only as an illustration of the effect of a hypothesis: that of continued exponential growth in the exploitation of mineral resources. The authors themselves rejected such a hypothesis as unrealistic and dedicated the rest of the book to developing better models. Indeed, it is clearly stated at page 63 of the 1972 edition of LTG that:

> Of course, the actual nonrenewable resource availability in the next few decades will be determined by factors much more complicated that can be expressed by either the simple static reserve index or the exponential reserve index. We have studied this problem with a detailed model that takes into account the many interrelationships among such factors as varying grades of ores, production costs, new mining technology, the elasticity of consumer demand, and substitution with other resources

So, accusing the LTG authors of "wrong predictions" on the basis of Table 4 is like accusing Euclid of not knowing geometry on the basis of one of his demonstrations based on the "reductio ad absurdum" method. If we read in detail the first LTG book, we find that none of the scenarios developed indicated the world should have "run out" of anything before the end of the twentieth century.

But Bailey's paragraph from his 1989 article had an extraordinary success and it spread in the press with the typical mechanism of urban legends. It even reached the scientific literature. In the section titled "Comments and Discussion" attached to a 1992 paper by William Nordhaus, we find a text by Robert Stavins, an economist from Harvard University, where we read:

> If we check today to see how the *Limits I* predictions have turned out, we learn that (according to their estimates) gold silver, mercury, zinc, and lead should be thoroughly exhausted, with natural gas running out within the next eight years. Of course, this has not happened.

It is obvious that Stavins is referring to Bailey's 1989 article, even though he does not cite it. Evidently, Stavins had not bothered to check whether his source was reliable. So, with this text, the legend of the wrong predictions was even reported in a peer-reviewed academic journal.

With the early 1990s, we can say that the dam gave way and a true flood of criticism swamped LTG and its authors. One after the other, scientists, journalists, and whoever felt entitled to discuss the subject, started repeating the same line over and over; the Club of Rome (or LTG) had predicted a catastrophe that did not take place and therefore the whole study was wrong.

There are innumerable examples for the spreading of this legend. For instance, here is a paragraph published by "The Economist" in 1997 (*The Economist*, 20 December 1997 p. 21).

> So, according to the Club of Rome, [petroleum] reserves should have been overdrawn by 50 billion barrels by 1990. In fact, by 1990 unexploited reserves amounted to 900 billion barrels -- not counting the tar shales, of which a single deposit in Alberta contains more than 550 billion barrels. The Club of Rome made similarly wrong predictions about natural gas, silver, tin, uranium, aluminum, copper, lead and zinc.

Here is another example from the popular book "The Skeptical Environmentalist" by Bjorn Lomborg (2001). At page 137 we find this paragraph which repeats nearly verbatim Bailey's 1989 text:

> Perhaps the most famous set of predictions came from the 1972 global best-seller "Limits to Growth" that claimed we would run out of most resources. Indeed, gold was predicted to run out in 1981, silver and mercury in 1985, and zinc in 1990.

The concept of wrong predictions made by LTG had become so widespread and entrenched that often it was not even necessary to provide details. For many people, LTG had been simply wrong and that was it. The late Nobel Prize winner Milton Friedman used the term "stupid projections" in an interview that Carla Ravaioli published in 1995 in her book "Economists and the Environment" (p. 33) (Ravaioli and Ekins 1995).

> *Friedman*: If we were living on the capital, the market price would go up. The price of truly limited resources will rise over time. The price of oil has not been rising, so we're not living on the capital. When that is no longer true, the price system will give a signal and the price of oil will go up. As always happens with a truly limited resource.
>
> *Ravaioli*: Of course the discovery of new oil wells has given the illusion of unlimited oil …
>
> *Friedman*: Why an illusion?

Ravaioli: Because we know it's a limited resource.

Friedman: Excuse me, it's not limited from an economic point of view. You have to separate the economic from the physical point of view. Many of the mistakes people make come from this. Like the stupid projections of the Club of Rome: they used a purely physical approach, without taking prices into account. There are many different sources of energy, some of which are too expensive to be exploited now. But if oil becomes scarce they will be exploited. But the market, which is fortunately capable of registering and using widely scattered knowledge and information from people all over the world, will take account of those changes.

The diffusion of these legends led sometimes to aggressive behavior. The LTG authors do not report having ever received the same kind of physical threats, or even death threats, that climate scientists started receiving after the "Climategate" case of 2009 (IPCC 2009; Fischer 2010). However, some statements that can be found over the Internet look worrisome, to say the least. For instance, Steven White (2002) suggested that the LTG authors who had made "wildly wrong" predictions "should be sent to the organ banks."

A list of the restatements of the legend of the "wrong predictions" would take a lot of space. It will suffice to say, here, that in December 2010, a Google search of the terms "Club of Rome," "wrong," and "predictions" produces about 17,000 results. Of these, a large majority report the legend as truth. Some of the restatements of the legend are very recent (e.g., Ball 2010) and some of these recent restatements appear in refereed journals (Popper et al. 2005; Radetzki 2010). Old legends, apparently, never die.

Within some limits, it is likely that the downfall of the LTG study was a natural consequence of the generally optimistic mood of the late 1980s regarding resource depletion. However, if we compare the story of LTG with some recent assessments of criticism against scientific results (McGarity and Wagner 2008; Hoggan 2009; Oreskes and Conway 2010), it is possible to surmise that active lobbying may have played a role in the formation of the strongly negative public opinion on LTG and it may have generated and diffused Bailey's successful critical lines (Bailey 1989).

The media debate normally involves the activity called "lobbying" in which special interest groups actively promote their interests with the public. There is nothing illegal in political lobbying, but serious problems arise when the lobbying action is obtained by means of covert action which often involve slandering and/or ad personam attacks, and especially when they are based on fabricated information – or at the least a failure to check facts. In these cases, these methods are called "spin tactics"; a form of propaganda aimed at manipulating the public opinion (Alasdair 2005; Gore 2007; Hoggan 2009; Oreskes and Conway 2010). We have no proof of active lobbying of this kind in the case of LTG, but it is possible to examine other known cases in order to identify common elements that may lead us to make reasonable hypotheses on this point.

One of earliest documented cases has been described by Kimm Groshong (2002) for the book "Silent Spring" by Rachel Carson (1962). In her book, Carson had criticized the overuse of DDT and other artificial pesticides and that went against the interest of a section of the chemical industry. As a consequence, the industry reacted with an aggressive public relation campaign. Quoting from Groshong's text:

> The minutes from a meeting of the Manufacturing Chemists' Association, Inc. on May 8, 1962, demonstrate this curious stance. Discussing the matter of what was printed in Carson's serialization in the *New Yorker*, the official notes read: The Association has the matter under serious consideration, and a meeting of the *Public Relations Committee* has been scheduled on August 10 to discuss measures which should be taken to bring the matter back to proper perspective in the eyes of the public. An Association program is being evolved, to be handled as a cooperative venture with other interested industry groups.

Groshong then goes on detailing how the Association selected seven points of criticism against "Silent Spring" and proceeded to diffuse these points in the press. This diffusion has been rather successful as, still in recent times, Rachel Carson is being accused of having "killed more people than the Nazi" because of her opposition to the overuse of the pesticide DDT (Oreskes and Conway 2010).

Of course, the right of the chemical industry to speak out on this matter is not in question. The problem is that the public was not made aware that many of the critical reviews of "Silent Spring" were the result of a concerted (and paid) effort specifically designed to influence the public's attitude.

The case of "Silent Spring" is possibly the first one known of an industrial lobby organizing a media campaign against a scientific study. A later, and better known, case is that of the health effects of tobacco. The discussion started in the 1960s (Diehl 1969) and raged for decades, reaching a peak in the 1980s. As part of a settlement agreed in 1998, the Tobacco Industry was forced by the US government to disclose all the documents related to the debate. Hence, we can document several examples of the methods used by the industry to discredit science and scientists. Such examples are described, for instance, by Thomas McGarity and Wendy Wagner in "Bending Science" (2008), by David Michaels in "Doubt is their product" (2008) and by Naomi Oreskes and Richard Conway in "Merchants of Doubt" (2010).

A typical example of the media campaign orchestrated by the tobacco lobby is the attack against the results published in 1981 by Takeshi Hirayama on passive smoke (Hirayama 1981). The results of Hirayama's study showed that the risk of contracting lung cancer for a nonsmoker was considerably increased by living in the vicinity of a smoker. As obvious, that was bad news for the tobacco industry and the reaction was rapid and aggressive. The industry attack was focused on alleged mistakes in the statistical treatment of the data in Hirayama's paper. In reality, all what the industry had was the opinion of some external experts that it was "possible" that Hirayama had made these mistakes, but that was sufficient to unleash a media blitz with the explicit purpose of "discrediting Hirayama's paper." As McGarity and Wagner (2008) report, the chairman of the US Tobacco Institute chose a moment in which Hirayama was away from his office, at a meeting, to send a telegram to the director of Hirayama's institute about the "very grave" error discovered in Hirayama's paper. In the report of McGarity and Wagner (2008), we also read the following description.

> Copies of the telegram were sent to Dr. Hirayama, the British Medical Journal and the press. A simultaneous press release announced that Dr. Mantel had "confirmed the exis-

tence of a fundamental mathematical error" in the Hirayama study. The institute's public relations department also distributed a package of the Mantel critique to a field force it had assembled with orders to "personally visit more than 200 news organizations nationwide" on Friday June 12. About one hundred video news clips were sent to local television stations and four hundred audio cassettes were sent to large radio stations.

The public relations blitz had its desired effect and news about Hirayama's "mistake" appeared in most of the media outlets in the United States. It has been said that about 80% of the US population was exposed to these news.

Hirayama was later able to demonstrate that he had been correct in his treatment of the data and his paper is considered today "a classic which has stood the test of time" (Ong and Glanz 2000). Nevertheless, the US Tobacco Institute had seriously damaged Hirayama's reputation and cast doubts on the value of his study that linger to this day.

There are many more examples of this kind of methods against scientific studies. Another interesting case is the so called "Nuclear Winter" (Turco et al. 1983) which described the results of a major nuclear war as producing a cooling effect that might easily destroy human civilization. The study came under attack from all the industrial and military sectors that were obviously damaged by these results. It was demonized to the point that it was listed in the magazine "Discovery" among "the 20 greatest scientific blunders of history" (Starr 2009). Later on, however, the results of the study were amply confirmed (Robock et al. 2007; Turco 2007; Starr 2009).

A further case of demolition of scientific results is the recent attack against climate science and climate scientists which has focused in particular against the so-called "hockey stick," the reconstruction of temperatures over the past 1,000 years (Mann et al. 1998). The basic soundness of the results was confirmed in later studies (e.g., National Academy of Sciences 2006). However, the study was subjected to a denigratory media campaign which includes today an entire book dedicated to its alleged shortcomings (Montford 2010). The debate rapidly degenerated in personal attacks against the principal author of the study, Michael Mann, who reported having received all sorts of insults and menaces, including death threats. The attacks were extended to Mann's colleagues and to climate science in general, culminating with the theft of data from the server of the Climate Research Unit of the University of East Anglia. The scandal that followed is known as "Climategate" (IPCC 2009; Fischer 2010) and it led to further demonization of climate science and of climate scientists.

From these examples, it is clear that there exists a pattern in these media spin campaigns. It consists in finding a weak point in the scientific study under attack and concentrating the media attention on that specific point, alone, neglecting all the evidence which, instead, supports the study. The "weak point" may be only described or perceived as such but, with sufficient media volume, it is possible to concentrate the attention of the public on marginal points and use these points as a leverage to demolish even well-established models and theories.

In the case of "The Limits to Growth," we clearly see the same technique at work: finding a single error, no matter how small and even nonexisting, and use it to discredit a whole study and its authors. This is what Ronald Bailey (1989) did when he picked up an early misinterpretation by other authors (Passel et al. 1972) to find that "single error" sufficient to discredit the whole study. The error that did not exist was not important for that purpose.

We have no document available that proves that, in some smoke filled room, representatives of the extractive industry gathered to decide what measures to take against the authors of LTG, as Groshong (2002) reports it happened for the case of Rachel Carson's "Silent Spring." What we can say for sure is that the denigration action against LTG was extremely successful because it played on some innate characteristics of the human mind; in particular to our tendency of believing what we want to believe and disbelieving what we do not want to believe. It is, in the end, the Cassandra Effect (and, of course, Cassandra was right).

So, with the demonization of LTG we lost several decades in which the methods and the results of LTG were pushed outside of the boundaries of political correctness and of scientifically sound ideas, as well as off the political and world stage. That may change in the future, but the damage done will not be easily remedied.

Chapter 11
The Return of World Modeling

The eclipse of "The Limits to Growth" lasted for a period that we may take as approximately corresponding to the 1990s. Apart from a brief flaring of the debate resulting from the publication of the second version of LTG (Meadows et al. 1992; Nordhaus 1992a), during this period the scientific debate on LTG was basically muted and very few studies appeared that would pursue the subject of world modeling in general. The LTG controversy had made scientists wary of entering an area where it was very hard – if not impossible – to get funding and where the results obtained were sure to generate opposition in the reviewing process of scientific journals. In the popular press, LTG was cited almost only in order to define it as a thoroughly flawed study; if not a purposeful attempt of deceiving the public. Even the environmental movement, in its various shades and versions, had focused on concepts such as "sustainable development" (United Nations 1987) that, under several respects, could be taken as opposed to the idea that there exist limits to growth.

Yet, the world is always changing and the transformations that took place with the end of the 1990s and the start of the 2000s led to a different view of the situation. If the 1990s had been the years of "The End of History" (Fukuyama 1992), the 9/11 attacks of 2001 showed that history had not ended at all. On the contrary, a new historical period was opening; one in which the "peace dividends" that some had seen as the result of the fall of the Soviet Union were to be used for a new round of war against new enemies.

The 1990s had also been the time of the "new economy," the idea that the Internet and information technology would usher a new era of prosperity for everyone who could set up a "dot-com" site. It was a moment of optimism not unlike the one generated by the novelty of nuclear energy in the 1950s. But the new century brought a fundamental disillusion on the capability of the economy to keep growing forever. The NASDAQ stock market badly crashed in 2000, putting an end to the dot-com bubble. The Dow-Jones index crashed in 2008, generating the collapse of the world's financial markets. Some saw the financial crisis that started in 2008 as similar to the one which ushered the "Great Depression" of 1929. This disillusion was enhanced

U. Bardi, *The Limits to Growth Revisited*, SpringerBriefs in Energy: Energy Analysis, DOI 10.1007/978-1-4419-9416-5_11, © Ugo Bardi 2011

by the second oil crisis that culminated with the burst in oil prices of 2008 which touched the all time high of almost 150 dollars per barrel.

All that was a reminder that the Cassandras of the 1970s, the authors of LTG, might not have been as wrong as most people had come to believe. Their emphasis on some basic problems: resource depletion, pollution, and overpopulation had not been misplaced. All three problems had been forgotten in the wave of optimism of the 1990s, but came back with a vengeance with the new century.

With the twenty-first century, interest in the future or mineral resources was spurred by renewed concerns about crude oil depletion. The reappraisal started in 1998, with a milestone paper published by Colin Campbell and Jean Laherrere on Scientific American. In that study, the authors took up again the model proposed by Marion King Hubbert (1956) to describe the world's crude oil production. They proposed that the worldwide peak of oil production ("peak oil") would occur at some moment during the first decade of the twenty-first century, most likely around 2005. Today, it is not possible to say with certainty whether the peak has actually occurred around that data; although it may be the case (e.g., Aleklett et al. 2010). What is certain is that the world's production of conventional oil has been static, or perhaps slightly declining, since 2005, while that of "all liquids" (combustible liquid fuels obtained by sources such as tar sands) has not been significantly increasing, on the average, from 2004. These data lend credence to the idea that the world will be soon facing a reduction in the availability of liquid fossil fuels.

Peak Oil has never been a mainstream theory, but it never was object of the political attacks that had reduced LTG to the level of fringe idea for lunatics. So, the diffusion of studies on oil depletion helped in changing the intellectual climate about mineral depletion in general; as shown by several studies which appeared on the subject (see, e.g., Sohn 2005; Tilton 2006; Gordon et al. 2006; Bardi and Pagani 2007; Bardi 2008a; Diederen 2010; Valero and Valero 2010; Bihouix and De Guillebon 2010; Rademaker and Korooshy 2010).

In recent times, fears related to minerals depletion have been heightened by the case of rare earths; minerals of great importance for the electronics industry which are supplied almost exclusively from mines located in the Republic of China (Bradsher 2010). Several other mineral resources have been found to have already peaked and to be declining in terms of production (Bardi and Pagani 2007). Depletion, actually, is not affecting just mineral resources, but also those which are normally considered as "renewable." For example, it has been reported that the loss rate of topsoil in Europe is on the average of 3–40 times faster than the value that would be sustainable (Verheijen et al. 2009). The loss of fertile soil brings an increasing need for artificial fertilizers that can only come from fossil fuels which are also facing a production crisis. Resource depletion, therefore, is a general problem that cannot pass unobserved any more.

Another area where the concepts expressed in LTG are gaining attention again is that of pollution. It is true that in the past there have been several successes in fighting local pollution, e.g., smog in urban areas. However, when dealing with "persistent pollution," the situation is different. In this case, the most important persistent pollutant is the excess of carbon dioxide, CO_2, in the atmosphere, the main cause of global warming. CO_2 concentration continues increasing, defeating all attempts to cap emissions generated by fossil fuels. With the manifestations of climate change becoming more and more visible in hurricanes, wildfires, heat waves, droughts, and floods (Hansen 2007; IPCC 2007; Martens et al. 2010). The fact that conventional crude oil may have peaked is not easing the climate problem (Garcia 2009; Bardi 2009), since it may direct the industry to extracting liquid fuels from "dirty" sources, such as tar sands and coal, which emit more CO_2 per unit energy produced.

Another important global issue which is returning under attention is population. Population control had become a political issue even before LTG; as evidenced by such texts as Hardin's paper on "The Tragedy of Commons" (1968) and by Ehrlich's book The Population Bomb (1968). However, the fate of population control policies followed the same parable of LTG. That is, after a phase of strong interest, the concept was demonized and disappeared from the political horizon, at least in Western Countries. Some authors (King and Elliot 1997) claimed that a phenomenon called "Hardinian Taboo" had stopped people from voicing their views on overpopulation for fear of stirring controversy.

Nevertheless, the population question never completely disappeared from the political horizon and it remained a live issue in those countries where the demographic transition was slow to arrive; such for China where the "one child per family" policy was started in 1979 and has been successful in slowing down population growth (Feeney and Feng 1993).

Today, there are signs that the population debate is being reopened, with opposite fears being voiced: overpopulation (e.g., Butler 2004) and population collapse (Pearce 2010; Connelly 2008). Paul and Anne Ehrlich (2009) have also reexamined "The Population Bomb" (Ehrlich and Ehrlich 1968) more than 40 years after its publication. They report that the book had several flaws, the main one of having underestimated the effect of the "Green Revolution" on agriculture. Nevertheless, according to the authors, the main theme of the book, that of a limited capacity of the ecosystem to produce food, remains valid.

It may be possible that the whole world is going through a demographic transition of the same kind as that experienced by affluent countries, as indicated by the results of the latest LTG study (Meadows et al. 2004). In other words, population growth may not be keeping the exponential rate that was so worrisome in the 1950s and 1960s and might stabilize or start declining by itself, without the need of active measures to control it. However, for the time being, growth continues and it is not obvious that food production will keep matching the world demand, especially in view of the growing concerns about water supply for irrigation (Hanjra and Qureshi 2010).

From these considerations, we see how concerns about issues of resources, climate, and population have rekindled the interest in the problem of the ultimate fate of our civilization. The collapse of ancient civilizations is a well-known event (e.g., Tainter 1988) and some successful books, such as Jared Diamond's "Collapse" (2005), brought the problem back to the attention of the public. That led to reexamine the LTG reports that started to be seen not any more as "Chicken Little with a Computer" (Meadows 1999a). This return of interest has resulted in a number of studies reevaluating the 1972 LTG book.

Already in 1995, Raymond Mikesell, an economist, had written a positive reappraisal of "The Limits to Growth." But the real change started in 2000 with a paper by Matthew Simmons titled "Revisiting The Limits to Growth: Could The Club of Rome Have Been Correct, After All?" Simmons (1943–2010) was an investment banker and an expert on crude oil who appeared frequently in the media. His paper on LTG is not the best known of his works but, surely, the popularity of the author made it very visible. Simmons was someone who would normally think "outside the box" and his paper is full of interesting insights; being also written in an easily readable style as – for instance:

> For a publication that is almost 30 years out of print, it is fascinating that anyone still even remembers what the book said. I have occasionally been privately amused at the passion this Club of Rome work still evokes. As I have heard this study thoroughly discredited, I have wondered whether the anger this book still creates is the equivalent of getting livid at a bartender "the morning after," when one's headache was so wicked.

Simmons's paper was the first of a series of positive reappraisals of "The Limits to Growth" that were published during the first decade of the 2000s. In 2004, Toby Gardner published a paper in "Resources policy" where he strongly criticized some rash dismissals of LTG, for instance by Bjorn Lomborg or Julian Simon. In 2008, the economist Peter A. Victor published his book "Managing without growth: slower by design, not disaster," Victor (2008) based in part on system dynamics scenarios inspired by "The Limits to Growth." Bardi (2008a) and Hall and Day (2009) both noted the relation of the LTG models with those related to "peak oil." Bardi also reviewed the story of the demonization of the LTG study (Bardi 2008b). A recent paper by Nørgård et al. (2010) has favorably reviewed the story of LTG while another by Eastin et al. (2010) makes the connection between the old debate on the LTG study and the present debate on climate change. The paper by Eastin et al. contains several imprecisions regarding the LTG story and seems to give credit to the existence of the alleged "errors" present in the 1972 book. Nevertheless, it makes a fundamental point in describing the policy problems surrounding the two cases; noting the enormous challenges involved in mitigating both resource depletion and climate change. Another attempt to link climate change with resource depletion as modeled in LTG was performed by Dolores Garcia in 2009; using the world3 model in order to estimate how the peaking and decline of fossil fuels would affect the emission of greenhouse gases.

Important steps in the reappraisal process were the papers by Hall and Day (2009) and by Graham Turner (2008) which analyzed in detail how the various LTG scenarios had compared with historical reality in the 30 years since the first book. The result, that will be surprising for many, is that not only the "base case" scenario is the one that has fared best but also that it has described with reasonable approximation the historical evolution of the world system up to now. According to Turner, population growth closely followed the calculated evolution and the historical data for agricultural production in terms of food per capita turned out to be in reasonable agreement with the base case scenario; within a factor of about 5%. The historical data for industrial production, too, turned out to be in acceptable agreement with the base case scenario, with a discrepancy of less than 15%. Finally, Turner examined an upper and a lower bound in the available estimates of mineral resources, finding that the lower bound corresponds well to the "base case" scenario, whereas the upper bound corresponds to more optimistic LTG scenarios which assume an important role of technology in the extraction process.

On the whole, these are excellent results considering the time span involved. The performance of the LTG "base case" scenario is all the more impressive considering that it was one of the first attempts to use system dynamics to model the world's economy and that it was based on a series of assumptions that, at the time, could not be easily verified. In any case, however, it may be appropriate to repeat here that LTG was not meant to be a forecasting tool. Reality and the "base case" scenario may well diverge in the future, but that would not undermine the validity of the study and of its methods.

So, the approach to world modeling of "The Limits to Growth" was not wrong, despite the legends that appeared in the late 1980s and which are still common today. Indeed, we are discovering that we have excellent methods to tell us something about the future. When we start from sound physical principles, reasonably good data, and a scientific approach, we can make good models and obtain useful advice from them.

There are several examples of good predictions in the past. For instance, in 1956, Marion King Hubbert correctly predicted the production trends for crude oil in the 48 US lower states. In climate science, starting from the early work of Tyndall (1863), in 1970 Laudadio had correctly predicted the concentration of carbon dioxide for the end of the century (Laudadio 1970). Later on, Wallace Broecker (1975) correctly said that "by early in the next century (carbon dioxide) will have driven the mean planetary temperature beyond the limits experienced during the last 1000 years."

Not that there have not been wrong predictions in the past or warnings that turned out to be misplaced. Of course, there were but, on the whole, an examination of the scientific literature shows that "wrong predictions" are much rarer than the common perception makes them to be, as it can be seen for the case of crude oil (Bentley 2000). It can also be seen that projections of future production of mineral commodities are more often too optimistic than the reverse (Sohn 2005). In other words, it is just another false legend that scientists are prone to fall to the "Chicken Little Syndrome," that is of believing on the basis of scant data that important resources such as minerals and food are soon going to run out.

So, we were warned with decades in advance about the problems we face nowadays. If we had listened, there would have been time to prepare for resource depletion and global warming already 30 years ago. However, warnings were ignored and the messengers demonized on the basis of misinterpretations and legends about "wrong predictions."

One of the consequences of the eclipse of "The Limits to Growth" in the 1990s was to slow down the development of studies on world modeling intended as including both the economy and the ecosystem as endogenous factors. Certainly, nearly all who examined this issue found that public funding to support their work essentially impossible to obtain. Today, an examination of the field (Fiddaman 2010a, b), shows that several such models have been developed or are being developed. However, the field does not seem to have reached the level of activity that one would expect for such an important issue. Climate modeling (IPCC 2007) appears to have reached a much more sophisticated level and to enjoy a larger financial support. However, the situation may change. It has been recently announced (Technology Review 2010; Fiddaman 2010a, b) that an ambitious European plan is in progress to build a model that will simulate the entire planet (including both the ecosystem and the socioeconomic system) on a scale and on a complexity never attempted before. This plan is based on upgrading a previous system called the "Earth Simulator," a very fast computer system designed mainly for climate studies. So far, it is just an announcement, but it is a clear indication of the renewed interest in world modeling of the kind pioneered by "The Limits to Growth."

So, we see a resurgence of world models which, in the future, may put back world modeling at the basis of the debate on the future of society. Whether that will allow us to solve the challenges we face remains to be seen, but at least it gives us a chance.

Chapter 12
Conclusion: The Challenges Ahead

The group of intellectuals who created the "Club of Rome" in 1968 had ambitious goals. They sponsored "The Limits to Growth" study with the idea that it would help them in understanding what they called the "world problematique" or "the predicament of mankind"; a set of crucial problems regarding the use and the distribution of the world's resources. They probably believed that, on the basis of the results of the LTG study, it would have been possible to convince the public and the world's leaders that it was necessary to gently steer the world's economic system towards a sustainable, steady state level, providing a reasonable level of material prosperity for everyone. But this idea turned out to be extremely difficult to put into practice.

The first problem encountered was the difficulty of having the message understood. In practice, the LTG study was widely misread and misinterpreted. Scenarios were mistaken for prophecies of doom, the need for concerted action was seen as a call for world dictatorship, the plea for equality taken as an attempt to destroy the lifestyle of people living in Western Countries. In addition, the concept that it was necessary to stop polluting activities and to slow down economic growth generated strong opposition from powerful industrial and political lobbies.

This stumbling block was not clearly identifiable at the time of the first LTG book, in 1972. The Club of Rome was aware of possible negative reactions to the study and had presented it to a number of independent reviewers before actual publication. The survey returned a certain perplexity on some issues but, on the whole, nothing that would prefigure the backlash that took place later. Evidently, the wide distribution of the study generated reaction mechanisms that were impossible to predict on small scale tests. In particular, in 1972, "spin" campaigns destined to demolish scientific results for the sake of special interests (Oreskes and Conway 2010; McGarity and Wagner 2008) were still relatively rare and not yet perfected. But, by the late 1980s, these techniques were well known and may have been utilized against the LTG study.

Whether it was as the result of a media campaign or simply as the result of the human tendency of disbelieving bad news (The "Cassandra Effect"), the LTG study was strongly criticized, then ridiculed, and finally consigned to the dustbin of wrong scientific theories, together with Lamarck's ideas on the evolution of the neck of giraffes[1] and Aristotele's statements on the spontaneous generation of maggots from dirt. Entrenched beliefs about the "wrong predictions" of LTG remain widespread today and are a major problem that appears whenever trying to reexamine the results of the study.

The difficulty in understanding the problems identified by the LTG study is compounded by the difficulty of putting into practice actions designed to solve them. Global problems require global action. The trouble is that global action, no matter of what kind, cannot be implemented simply by exhortations and good will. It needs political mechanisms and functioning institutions.

This is a further stumbling block that was not easily identifiable when the first LTG book appeared in 1972. The Club of Rome and the LTG authors often stated the need for concerted action at the worldwide level for solving – or at least mitigating – the world's problems. But they do not seem to have ever stated what specific mechanisms they saw as suitable for obtaining the desired results. In practice, the general consensus on this matter seems to have been – and to be still today – that global actions designed to fight environmental threats can be obtained by means of international treaties. But this line of action, too, proved to be difficult: in particular for the most important problems: those related to the management of global resources.

There exist today several cases of successful international treaties designed to fight environmental threats. A classic case, discussed at length in the 2004 version of LTG, is that of the banning of chlorofluorocarbons (CFCs); compounds that cause the depletion of the stratospheric ozone layer that protects living beings from ultraviolet radiation. The treaty against CFCs, the "Montreal Protocol," was signed in 1987 and it has been successful in reducing the use of CFCs and in starting a slow return of the ozone layer to pre-CFC levels. As other examples, one may consider the 1963 treaty banning nuclear explosions in the atmosphere, the 1997 Ottawa treaty against antipersonnel mines, and various treaties limiting overfishing.

Unfortunately, these treaties deal with specific and relatively simple problems. Acting on more complex – and more dangerous – problems turned out to be much more difficult. Regarding global warming, for instance, we see that the world institutions are moving very slowly – if at all – in dealing with a problem which is, potentially, the most dangerous ever faced by humankind. The Kyoto protocol, devised in 1997, was a step forward, but it did not succeed in reducing the emissions of CO_2 in the atmosphere from industrial activities. Then, the failure of the Copenhagen meeting in 2009 was a major stumbling block in the attempt for controlling global warming. It looks extremely unlikely, at present, that it will be possible to agree on a new treaty that might do better than the Kyoto one, or even simply replace it, before it expires in 2012.

We do not seem to be able to do any better in other critical environmental areas. It is possible that some countries are limiting the exploitation of local mineral reserves in order to extend their duration (Reynolds 2000). But, at the global level, the only proposed treaty in this sense has been the "Oil Protocol" devised by Colin

---

[1] Actually, Lamarck never mentioned the neck of giraffes in the terms normally attributed to him (Ghiselin 1994). It is just another false legend related to a scientific theory.

Campbell in 1996 (at that time known as the "Rimini Protocol") (Heinberg 2006). However, after some initial interest, the proposal was ignored.

The problem seems to lie with issues which require widespread changes in everyone's behavior. Banning CFCs was relatively easy: substitutes were readily at hand and most people did not notice that their new refrigerator was using a different fluid than the old one. But when it comes to such things as a "carbon tax" designed to fight global warming or to slow down oil depletion, everyone notices that some goods, from automotive fuels to home heating, go up in price. At the same time, companies dealing with these goods see their sales, and hence their profits, reduced. In the end, it becomes exceedingly difficult for politicians to propose and implement measures that reduce their popularity and damage some sectors of the economy, despite the benefits for society as a whole. Prins et al. (2010) define global warming as a "wicked" problem, a definition that may apply to resource depletion as well. Solving such problems involves radical changes in the structure of society and that makes them politically unsolvable and, in some cases, unmentionable.

If we want to implement measures that are unpopular and resisted by special interest lobbies, we must reach a level of shared vision that makes it clear that such measures are absolutely necessary. It is not impossible: in emergency situations, such as in wars or natural disasters, people understand the need of personal sacrifices for the sake of the common good. Given this situation, we might have to wait for a catastrophe of some kind to occur to awaken public consciousness (see, e.g., Baranzini et al. 2003). But, if we have to wait for the societal collapse envisaged by the LTG scenarios to be absolutely evident, it will be too late to do something to prevent it by means of international treaties.

But, if we cannot implement wide sweeping, global actions, we can at least do our best to mitigate the negative effects of the problems that the LTG study had already identified in 1972: resource depletion, persistent pollution, and soil erosion; all symptoms of the "overshoot" conditions in which the human society finds itself. Considering that the perfect is the enemy of the good, we may consider a policy of small steps which are still better than nothing and have the advantage of being feasible. "Small steps" in this context means pushing for measures which are flexible and which can be progressively adapted to changes in the extent of the problems and in the public understanding of the situation. For instance, a carbon tax can only be as high as it is politically acceptable (Prinse et al. 2010) and such level may not be high enough to pay for the so called "externalities" or "external costs" involved with CO_2 emissions. But even a small carbon tax may be better than no tax at all.

Then, we need to make sure that the small steps that may be taken are consistent with the desired goals. It is perfectly possible to understand what the problems are and still take decisions that obtain the opposite effect as the one wanted (Meadows 1999b). In order to take steps in the right direction, we need to know where we are going; and that can be obtained only by means of models.

Many people seem to see models – of all kinds – as entities full of sound and fury and signifying nothing; little more than the dew of Basho's poems or the *Vanitas Vanitatum* of the Ecclesiastes. What can we learn about the real world, so incredibly complex and variegated, from equations coded into a computer memory?

But, all (or almost all) models can be useful if one understands their limits and their purposes. Models are attempts to build something that reflects the way we see the world according to the data we have. This point was clearly stated by the authors of the third edition of LTG as: (p. 133)

> For this model to be of any use, we will have to compare it to the "real world," but neither we nor you, our readers, have one agreed-upon "real world" to compare it to. All of us have only our mental models of the entity that is normally called the real world. Mental models of the surrounding world are informed by objective evidence and subjective experience. <..> What we mean by "*real world*" or "*reality*" is just the shared mental model of the authors of this book. The word *reality* can never mean anything more than the mental model of the user of that word.

The discrepancy between our mental models and the real world may be a major problem of our times; especially in view of the difficulty of collecting, analyzing, and making sense of the unbelievable amount of data to which we have access today. Unfortunately, most people – including policy makers – seem to trust more the scraps and bits of news they hear in TV or read in the press than the results of expensive and time consuming modeling efforts.

In the end, a model is a tool for taking decisions and any decision taken is the result of a process of reasoning that takes place within the limits of the human mind. So, models have eventually to be understood in such a way that at least some layer of the process of simulation is comprehensible by the human mind. Otherwise, we may find ourselves acting on the basis of models that we don't understand, or no model at all. Seymour Papert in his "Mindstorms" (1980) had correctly understood the situation when he spoke of "mind sized models."

No model can change beliefs which are deeply entrenched inside the human mind. But, there is some hope that mind sized models may help people overcome their biases, at least in part. For this purpose, quantitative world models may not be an absolute necessity. Dynamic ways of thinking based on concepts such as "Leverage points" (Meadows 1999b) and "Misperception of Dynamic Systems" (Moxnes 2000, 2004, Moxnes and Saysel 2009) can help us avoiding the trap of overexploitation. The use of dynamic concepts can help managing everyday life in business (Sterman 2000) and in social issues (Madden 1979). The concept of "The Fifth Discipline," management based on dynamic concepts, has been described by Peter Senge (Senge 1990, Senge and Sterman 1992).

A model can amplify one's mental power by providing methods for defining and testing what are the variables that move the world. In other words, a good model can overcome the very poor models on which most decisions – at all levels – seem to be taken today in the world. This is, perhaps, the best hope we have to meet the enormous challenges we are facing. That was, and still is, the promise of LTG that started in 1972 in an intellectual journey that lasts till date.

About the Author

Ugo Bardi teaches Physical Chemistry at the University of Firenze in Italy. He is engaged in research on materials for new sources of energy and on studying the dynamic exploitation of natural resources. He is a member of the scientific committee of the Association for the Study of Peak Oil (ASPO) and president of the Italian section of the same association. He is also a regular contributor of "The *Oil Drum*" site as well as a contributor of several other sites dedicated to renewable energy and sustainability.

U. Bardi, The Limits to Growth Revisited, SpringerBriefs in Energy: Energy Analysis,   105
DOI 10.1007/978-1-4419-9416-5, © Ugo Bardi 2011

References

Ackoff, R.L., 1971. "Towards a System of Systems" *Management Science*, Vol 17, No 11, pp. 661–670

Alasdair, R.S., 2005. "Spin Control and Freedom of Information: Lessons for the United Kingdom from Canada". Public Administration 83: 1. doi:10.1111/j.0033-3298.2005.00435.x

Aleklett, K., Höök, M., Jakobsson, K., Lardelli, M., Snowden, S., Söderbergh, S. 2010. "The Peak of the Oil Age" *Energy Policy*, Volume 38, Issue 3, Pages 1398–1414

Allen, R.G.D., 1955. "The Engineer's Approach to Economic Models" *Economica*, New Series, Vol. 22, No. 86, pp. 158–168

Ashford, Oliver, M., 1985. Prophet or Professor? Life and Work of Lewis Fry Richardson. Bristol: Adam Hilger. ISBN 978–0852747742

Atkinson, Nancy, 2010. "New Dates for Final Shuttle Launches" Universetoday, http://www.universetoday.com/67524/new-dates-for-final-shuttle-launches/ (accessed Sep 11 2010)

Ayres, Robert, U., 1998. Technological Progress: A Proposed Measure, *Technological Forecasting and Social Change*, Volume 59, Issue 3, pp 213–233

Ayres, Robert, U., 2001. "The minimum complexity of endogenous growth models: the role of physical resource flows" *Energy*, 26, 9, 817–838

Ayres, R., 2006. "Lecture 5: Economic growth (and Cheap Oil)" http://www.cge.uevora.pt/aspo2005/abscom/ASPO2005_Ayres.pdf

Ayres, R., 2007. "On the practical limits to substitution" Ecological Economics vol. 61, p. 115–128

Bailey, Ronald, 1989. "Dr. Doom" Forbes, Oct 16, p. 45

Bailey, Ronald, 1993. Eco-Scam: The False Prophets of Ecological Apocalypse, St. Martins, Washington

Bak, Per, 1996. "How Nature Works", Copernicus, New York

Baker, C.S., and Clapham, P.J., 2004, Trends in Ecology and Evolution Vol. 19 No.7, p.1

Ball, Tim, 2010. "Disastrous Computer Models Predictions From Limits to Growth to Global Warming" Canadian Free Press, http://www.canadafreepress.com/index.php/article/22444 accessed 22 Nov 2010

Baranzini, Andrea, Chesney, Marc, Morisset, Jacques, 2003. "The impact of possible climate catastrophes on global warming policy" *Energy Policy*, 31 pp. 691–701

Bardi, U., 2005. "The mineral economy: a model for the shape of oil production curves," *Energy Policy* vol. 33 pp. 53–61

Bardi, U., 2007a. "Peak Oil's ancestor: the peak of British Coal Production in the 1920s" ASPO Newsletter n. 73. http://www.energiekrise.de/e/aspo_news/aspo/newsletter073.pdf, (accessed Sep 05 2010)

Bardi, U., 2007b. "Energy Prices and Resource Depletion: Lessons from the Case of Whaling in the Nineteenth Century" *Energy Sources, Part B: Economics, Planning, and Policy*, Volume 2, Issue 3, pages 297–304

Bardi, U., and Pagani M., 2007. "Peak Minerals", The Oil Drum, http://europe.theoildrum.com/node/3086 (accessed August 31 2010)

Bardi, U., 2008a. "Peak Oil and "The Limits to Growth": two parallel stories", The Oil Drum, Feb 16; http://www.theoildrum.com/node/3550 (accessed July 20 2010)

Bardi, U., 2008b. "How 'The Limits to Growth' was demonized." The Oil Drum, March 9, http://www.theoildrum.com/node/3551 (accessed July 20 2010)

Bardi, U., 2008c. "The Universal Mining Machine" The Oil Drum, www.theoildrum.com/node/3451 (accessed August 31 2010)

Bardi, U., 2009. "Fire or Ice? The role of peak fossil fuels in climate change scenarios" The Oil Drum, http://europe.theoildrum.com/node/5084, (accessed February 2 2011)

Bardi, Ugo, Lavacchi, Alessandro, 2009. "A Simple Interpretation of Hubbert's Model of Resource Exploitation." *Energies 2*, no. 3: 646–661

Bardi, U., 2010. "Extracting minerals from seawater. An energy analysis". Sustainability 2(4), 980–992; doi:10.3390/su2040980

Barney, G.O., 1980. "The Global 2000 Report to the President of the U.S." Pergamon Policy Studies, New York

Beckerman, W., 1974. "Economists, Scientists and Environmental Catastrophe" Oxford Economic Papers. 24, p327

Bentley, R.W., 2000. "Were those past oil forecasts really so wrong?" In: The industry newsletter: To-morrow's Oil. OneOffshore, part of the PetroData Group, Aberdeen, November

Bihouix Phlippe and De Guillebon, Benoit, 2010. "Quel futur pour les métaux?" EDP Sciences, Paris

Bongaarts, J.P., 1973. "A review of the population sector in The Limits to Growth", *Studies in Family Planning*. Vol 4, No. 12, pp 327–334

Boyd, R., 1972. World dynamics: a note. *Science* 177:516–519

Bradsher, K., 2010. "China Said to Widen Its Embargo of Minerals" New York Times, Oct 19, http://www.nytimes.com/2010/10/20/business/global/20rare.html, (accessed Nov 01 2010)

Broderick, D., 2001. "The Spike", Tom Doherty Associates, New York

Broecker, W.S., 1975. Climatic Change: Are We on the Brink of a Pronounced Global Warming? *Science*, New Series, Vol. 189, No. 4201 (Aug. 8, 1975), pp. 460–463

Brown, L.S., Castillejo, L., Jones, H.F., Kibble, T.W.B., and Rowan-Robinson, M., 1973. "Are there real limits to growth? A reply to Beckerman" *Oxford Economic Papers*, new Series, Vol 25, N. 3, 455–460

Brown, M.T., 2004. "A picture is worth a thousand words: energy systems language and simulation", *Ecological Modelling*, 178: 83–100

Burdekin, R., 1979. "A dynamic spatial urban model: A generalization of Forrester's urban dynamics model" *Urban Systems*, Volume 4, Issue 2, 1979, Pages 93–120

Busby, S., 2006. "Pure Charity" http://www.after-oil.co.uk/charity.htm (accessed 04 Sep 10)

Butler, Colin, D., 2004. "Human Carrying Capacity and Human Health " *PLoS Med*. 1(3): e55

Caffrey, M., 2000 "Toward a History-Based Doctrine for Wargaming." Aerospace Power Journal, Fall 2000. http://www.airpower.maxwell.af.mil/airchronicles/cc/caffrey.html

Campbell, Colin, J., and Laherrère, Jean, H., 1998 "The End of Cheap Oil" *Scientific American*, March issue. pp. 78–83

Carson, Rachel, 1964. Silent Spring. Boston: Houghton Mifflin

Catton William, R., 1982. "Overshoot", University of Illinois press, Champaign. ISBN 0-252-009826

Club of Rome, 1970. "The Predicament of Mankind" http://sunsite.utk.edu/FINS/loversofdemocracy/Predicament.PTI.pdf (accessed Oct 25 2010)

Cole, H.S.D., Freeman, C., Jahoda, M., Pavitt, K.L.R., 1973. "Models of Doom" Universe Books, New York

Connelly, Matthew, 2008. "Fatal Misconception", Harvard University Press, Cambridge

Coyle, R.G., The technical elements of the system dynamics approach, *European Journal of Operational Research*, Volume 14, Issue 4, December 1983, Pages 359–370, ISSN 0377–2217, DOI: 10.1016/0377-2217(83)90236-9

Daly, Herman, 1977. "Steady State Economics", Island Press, Washington, ISBN: 155963071X

Daly, Herman, E., 1997. "Georgescu-Roegen versus Solow/Stiglitz" *Ecological Economics* vol 22 pp 261–266

De Wit, M., 2005. "Valuing copper mined from ore deposits" *Ecological Economics*, vol. 55 pp. 437–443

Deffeyes, K.S., 2001. "Hubbert's Peak: The Impending World Oil Shortage", Princeton: Princeton University Press

Deffeyes, K.S., 2005. "Beyond Oil" ISBN −13 978-0-8090-2956-3

Diederen, A.M., 2010. "Managed Austerity and the Elements of Hope" Eburon, Delft, ISBN 9789059724259

Diehl, H., Tobacco and Your Health, New York: McGrawHill, 1969

Dyson, F.J., 1960. "Search for Artificial Stellar Sources of Infra-Red Radiation". *Science* 131 (3414): 1667–1668

Eastin, John, Grundmann, Reiner; Prakash, Aseem, 2010 "The two limits debates: 'Limits to Growth' and climate change", Futures, In Press

Ehrlich, Paul, R., and Ehrlich, A., 1968. "The Population Bomb". New York: Sierra Club–Ballantine Books

Ehrlich, Paul, R., and Ehrlich, A., 2009. "The Population Bomb Revisited" http://www.population-media.org/wp-content/uploads/2009/07/Population-Bomb-Revisited-Paul-Ehrlich-20096.pdf. (accessed Nov 6th 2010)

Ettinger, Robert, 1974. Man into Superman (HTML), Avon. ISBN 0-380-00047-4

Feeney, G., and Feng, W., 1993. "Parity progression and birth intervals in China: The influence of previous policy in hastening fertility decline." Population and Development Review 19, 1, pp. 61–101

Fiddaman, T., 2010a. "System Dynamics World Library" http://www.metasd.com/models/ (accessed Nov 06 2010)

Fiddaman, T., 2010b. "The model that ate Europe" http://blog.metasd.com/2010/05/the-model-that-ate-europe/ (accessed Nov 07 2010)

Fischer, D., 2010. "Climategate Scientist Cleared in Inquiry, Again" Scientific American, July 1st, http://www.scientificamerican.com/article.cfm?id=climategate-scientist-cleared-in-inquiry-again (accessed Oct 29 2010)

Forrester, J., W., 1951. U.S. Patent 2,736,880 "Multicoordinate digital information storage device" (coincident-current system), filed May 1951, issued Feb 28, 1956

Forrester, J.W., 1958. Industrial Dynamics: A Major Breakthrough for Decision Makers. Harvard Business Review, 36(4), July/August, 37–66

Forrester, J.W., 1961. Industrial Dynamics. Waltham, MA: Pegasus Communications. 464 pp

Forrester, J.W., 1969. Urban Dynamics. Waltham, MA: Pegasus Communications. 285 pp

Forrester, J.W., 1971. World Dynamics. Wright-Allen, Cambridge, MA

Forrester, J.W., Low G.W., Mass N.J., 1974. "The Debate on World Dynamics: A Response to Nordhaus", Policy *Sciences* 5, 169–190

Forrester, J.W., 1992. "From the Ranch to System Dynamics: An Autobiography" in "Management Laureates: A Collection of Autobiographical Essays", Vol. 1 edited by Arthur G. Bedeian, JAI Press, http://www.friends-partners.org/GLOSAS/Peace%20Gaming/System%20Dynamics/Forrester's%20papers/Forrester-Ranch.html

Forrester, J.W., 1998. "Designing the Future" paper presented at Universidad de Sevilla, Sevilla, Spain December 15, 1998 http://sysdyn.clexchange.org/sdep/papers/Designjf.pdf

Franses, P.H., 2002. "A concise Introduction to Econometrics", Cambridge University Press, London

Fukuyama, F., 1992. "The End of History and the Last Man". Free Press, New York, 1992. ISBN 0-02-910975-2

Gardner, Toby, 2004. 'Limits to Growth? A Perspective on the Perpetual Debate', *Journal of Integrative Environmental Sciences*, 1, 2, 121–138

Garcia Dolores, 2009. "A new world model including energy and climate change data", The Oil Drum, http://europe.theoildrum.com/node/5145, (accessed Aug 15 2010)

Georgescu-Roegen, Nicholas, 1971. "The Entropy Law and the Economic Process", Harvard University Press, Cambridge, MA

Georgescu-Roegen, N., 1977. The steady state and ecological salvation: a thermodynamic analysis. BioScience 27, 4, pp. 266–270

Georgescu-Roegen, N., 1979a. Myths about energy and matter. Growth and Change 10, 1, pp. 16–23

Georgescu-Roegen, Nicholas, 1979b. Comments on the papers by Daly and Stiglitz. In: Smith, V. Kerry (Ed.), Scarcity and Growth Reconsidered, RFF and John Hopkins Press, Baltimore, MD

Giarini, Orio, and Loubergé, Henry, 1978. "Les Rendements Decroissants de la technologie", Arnoldo Mondadori., Milano

Gibbon, E., 1776–1788. "The Decline and Fall of the Roman Empire" [William] Strahan and [Thomas] Cadell, in the Strand, London

Goeller, H.E., and Weinberg, Alvin, M. The Age of Substitutability. The American Economic Review, Vol. 68, No. 6 (Dec., 1978), pp. 1–11

Golub, R., Townsend, J., 1977. "Malthus, Multinationals and the Club of Rome" *Social Studies of Science* vol 7, p 201–222

Goodman, George, J.W., (under the pseudonym of "Adam Smith") 1987. "Debt, the Grim Reaper" The New York Times. April 12, http://www.nytimes.com/1987/04/12/books/debt-the-grim-reaper.html. (accessed Dec 2, 2010)

Gore, A. 2007. "The Assault on Reason", Penguin, New York

Gordon, R.B., Bertram, M., Graedel, T.E., 2006. "Metal stocks and sustainability". *Proc. Natl. Acad. USA* 103(5), 1209–1214

Groshong, K., 2002. The Noisy Response to Silent Spring: Placing Rachel Carson's Work in Context. Pomona College, Science, Technology, and Society Department Senior Thesis http://www.sts.pomona.edu/ThesisSTS.pdf (accessed Sep 2009), not available any more in 2011; kindly provided by Richard Worthington (RKW14747@pomona.edu)

Hagens Nate, 2007. "Climate Change, Sabre Tooth Tigers and Devaluing the Future" The Oil Drum, http://www.theoildrum.com/node/2243 (accessed Feb 15 2011)

Hall, C.A.S., 1988. "An assessment of several of the historically most influential theoretical models used in ecology and of the data provided in their support" *Ecological Modelling*, vol. 43 p. 5–31

Hall, C.A.S., Powers, R., and W., Schoenberg, 2008. "Peak oil, EROI, investments and the economy in an uncertain future". Pp. 113–136 in Pimentel, David. (ed). Renewable Energy Systems: Environmental and Energetic Issues. Elsevier, London

Hall, Charles, A., and Day, John. W., Jr., 2009. "Revisiting the Limits to Growth after peak oil", *American scientist*, vol 97, p. 230

Hanjra, M.A., Qureshi, M.E., 2010. "Global water crisis and future food security in an era of climate change" *Food Policy*, Volume 35, Issue 5, Pages 365–377

Hardin, Garrett, 1968. "The Tragedy of the Commons," *Science*, vol. 162, pp 1243–1248

Hansen, James, 2007. "Climate catastrophe" *The New Scientist*, Volume 195, Issue 2614, Pages 30–34

Harte, J., Socolow, R.H., 1971. "Patient Earth", Holt, Rinehart and Winsto, New York

Heinberg, Richard, 2006. "The Oil Depletion Protocol" New Society Publishers, Canada, ISBN: 0865715637

Heinberg, Richard, 2010. "Beyond the Limits to Growth" Post Carbon Institute, Founation series concepts. Jul 28 issue. http://www.postcarbon.org/report/122404-foundation-concepts-beyond-the-limits-to (accessed 06 Sep 2010)

Heinlein, R.A., 1952. "Pandora's box" February 1952 issue of Galaxy magazine, pp. 13–22

Herrera, A., 1977. "Catastrofe o Nueva Sociedad? Modelo Mundial Latinoamericano". Ottawa, International Development Research Center

Hirayama, T., British Medical Journal 282(1981)1393

Hogan, G, 2006. Discover Oct 2006 http://www.discover.com/issues/oct-06/cover/

Hoggan, J., 2009. "Climate cover-up". Greystone, Vancouver

Hook, Mikael, Bardi, Ugo, Feng, Lianyong, Pang, Xiongqi, 2010. "Development of oil formation theories and their importance for peak oil". Marine and Petroleum Geology, 27, 9, 1995–2004, doi: DOI: 10.1016/j.marpetgeo.2010.06.005

Hotelling, Harold, 1931. The Economics of Exhaustible Resources. *Journal of Political Economy* 39: 137–75

Houthakker, Hendrik, S., 2002. "Are Minerals Exhaustible?" *The Quarterly Review of Economics and Finance*, vol. 42 p. 417–421

Hubbert, M.K., 1956. Nuclear Energy and the Fossil Fuels. Presented before the Spring Meeting of the Southern District, American Petroleum Institute, Plaza Hotel, San Antonio, TX, March 7–9, 1956

Hubbert, M.K., 1977. "Role of Geology in Transition to a Mature Industrial Society", *Geologische Runday* 66, n. 3, 654–678

Huebner, Jonathan, 2005. Technology Forecasting & Social Change, n. 72, p 980

IPCC, 2007. "Climate Change 2007: Report of the Intergovernmental Panel on Climate Change". Cambridge University Press, Cambridge, UK, 996 pp

IPCC, 2009. "Statement by Working Group I of the Intergovernmental Panel on Climate Change on stolen emails from the Climatic Research Unit at the University of East Anglia, United Kingdom". http://www.ipcc.ch/pdf/presentations/WGIstatement04122009.pdf. (accessed Sep 12 2010)

Jevons, William Stanley, 1866. "The Coal Question, An Inquiry Concerning the Progress of the Nation, and the Probable Exhaustion of Our Coal-Mines." Library of Economics and Liberty. http://www.econlib.org/library/YPDBooks/Jevons/jvnCQ.html (accessed Nov 28 2010)

Karnani, Mahesh and Annila, Arto, 2009. "Gaia, again" *BioSystems* vol 95 pp. 82–87

Kelly, K., 1994. "Out of Control" Fourth Estate. London

Kenney, J.F., 1996. Considerations about recent predictions of impending shortages of petroleum evaluated from the perspective of modern petroleum science. *Energy World* 240, 16–18

Kenney, J.F., Shnyukov, A.Y.E., Krayushkin, V.A., Karpov, V.G., Kutcherov, V.G.,Plotnikova, I.N., 2001. Dismissal of the claims of a biological connection for natural petroleum. *Energia* 22 (3), 26–34

King, M., Elliot, C., 1997. "A Martian View of the Hardinian Taboo". *British Medical Journal*, vol 315 pp. 1441–1443

Kirby, M. W., 1977. "The British Coalmining Industry, 1870–1946", MacMillan, London

Koehler, J.E., 1973. *The Journal of Politics*, 35 p 513–514

Kolata, G., 2009. Playing it safe in cancer research. The New York Times, June 28

Krugman, Paul, 2008. "Limits to Growth and Related Stuff" – The New York Times, http://krugman. blogs.nytimes.com/2008/04/22/limits-to-growth-and-related-stuff/ (accessed Dec 2, 2010)

Kurzveil, Raymond, 2005, "The Singularity Is Near: When Humans Transcend Biology" (Viking Books, New York. ISBN 0-670-03384-7)

Laherrere, Jean, 2004. "Present and future energy problems". HEC MBA Sustainable Development seminar, http://www.hubbertpeak.com/laherrere/HEC-long.pdf (accessed Nov 08 2010)

Larouche, L., H., 1983. "There are no Limits to Growth" The New Benjamin Franklin House, New York

Lasky, S.G., 1950. "How Tonnage and Grade Relations Help Predict Ore Reserves." *Engineering and Mining Journal* 151:81–85

Laudadio, L., 1970. "On the dynamics of air pollution: a correct interpretation", *Canadian Journal of Economics*, Vol 3, pp 563–571

Lewis, G., 2002. "Systems Thinking about Purpose", *Australasian Journal of Information Systems*, December special issue, pp 50–58

Lightfoot, H.D., 2007. "Understand the three different scales for measuring primary energy and avoid errors", *Energy* vol 32 pp. 1478–1483

Lomborg, Bjorn, 2001. "The Skeptical Environmentalist" Cambridge University Press, Cambridge, UK

Lotka, A.J., 1925. "Elements of physical biology"; Williams & Wilkins Co: Baltimore, MD

Lorenz, Edward N. (March 1963). "Deterministic Nonperiodic Flow". *Journal of the Atmospheric Sciences* 20 (2): 130–141

Lovelock, J.E., 1965. "A physical basis for life detection experiments". Nature 207 (7): 568–570. doi:10.1038/207568a0 DOI:dx.doi.org

Madden, M., 1979. "Recent Developments in Urban Dynamics: A Review Article " *The Town Planning Review*, Vol. 50, No. 2 , pp. 216–231

Mann, Michael, E., Bradley, Raymond S., Hughes, Malcolm K., 1998. "Global-scale temperature patterns and climate forcing over the past six centuries". Nature 392: 779–787. doi:10.1038/33859

Marxsen, Craig, S., 2003. The Independent Review, v. VII, n. 3, Winter 2003, pp. 325–342

Maxwell, James Clerk, 1868. "On Governors". 16. Proceedings of the Royal Society of London. pp. 270–283

McCleary, G.F., 1953. "The Malthusian population theory" Faber and Faber Limited, London

McGarity, T.O., Wagner, W.E., "Bending Science" Harvard University Press, Cambridge, MA, 2008

Meadows, Donella, H., Dennis, L., Meadows, Jorgen Randers, and William W., Behrens, III., 1972. "The Limits to Growth". New York: Potomac

Meadows, Dennis, William, W., Behren III, Donella Meadows, Roger F., Naill, Jorgen Randers, and Erich K.O., Zahn. 1974. Dynamics of Growth in a Finite World. Cambridge: Wright-Allen Press

Meadows, D., Richardson, J., Bruckmann, G., 1982. "Groping in the Dark: the first decade of global modelling." Wiley, Chichester

Meadows, Donella, H., 1999a. "Chicken Little, Cassandra, and the Real Wolf" Whole Earth, Spring 1999, http://wholeearth.com/issue/2096/article/72/chicken.little.cassandra.and.the.real.wolf (accessed August 20 2010)

Meadows, Donella, H., 1999b. "Leverage points: places to intervene in a system" The Sustainability Institute. www.sustainer.org/pubs/Leverage_Points.pdf (accessed Sep 14 2010)

Meadows, Donella H., Meadows, Dennis L. and Randers, Jorgen, 1992. Beyond the Limits. Chelsea Green, Post Mills, VT

Meadows, D.H., Randers, J., Meadows, D.L., 2004. Limits to Growth: The 30-Year Update Chelsea Green, White River Junction, VT

Mesarovic, M., and Pestel, E., Mankind at the Turning Point: The Second Report to The Club of Rome 1974. Dutton, New York, ISBN 0-525-03945-7

Michaels David, 2008. "Doubt is their product" Oxford University Press, New York

Mikesell, R.F., 1995. "The Limits to Growth, a reappraisal" Resources Policy, Volume 21, Issue 2, June 1995, Pages 127–131

Montford, A.W., 2010. "The Hockey Stick Illusion" Stacey International, London.

Moxnes, E., 2000. "Not only the tragedy of the commons: misperceptions of feedback and policies for sustainable development." System Dynamics Review 16(4), 325–348

Moxnes, E., 2004. "Misperceptions of basic dynamics, the case of renewable resource management." System Dynamics Review 20 (2) 139–162

Moxnes, E., and Saysel, A.K., 2009. "Misperceptions of global climate change: information policies." Climatic Change 93(1–2):15–37

Mukherje, Siddharta, 2010. "The Emperor of All Maladies: A Biography of Cancer" Scribner, New York

Myrtveit, Magne, 2005. "Working papers in System dynamics" -http://www.ifi.uib.no/sd/working-papers/WPSD1.05WorldControversy.pdf (accessed July 20, 2010)

National Academy of Sciences, 2006. Surface Temperature Reconstructions for the Last 2,000 Years, http://books.nap.edu/catalog.php?record_id=11676#toc (accessed July 2010)

Nebbia, G., 1997. "Bisogno di Storia e di Futuro" Futuribili, New Series, Gorizia (Italy) 4(3) 149–82

Nebbia, G., 2001. "Twenty twenty five" Futures, 33 p 43–54

Nordhaus, W.D., 1973. "Word Dynamics: Measurements without Data", The Economic Journal n. 332

Nordhaus, W.D., 1992a. "Lethal Models" Brookings Papers on Economic Activity 2, 1

Nordhaus, W.D., 1992b. "An optimal transition path for controlling greenhouse gases" Science, 258(20 November) 1315–1319

Nordmann, A., 2008. "Singular Simplicity The story of the Singularity is sweeping, dramatic, simple – and wrong" IEE Spectrum, http://spectrum.ieee.org/robotics/robotics-software/singular-simplicity, (accessed Aug 2010)

Nørgård, Jørgen Stig, Peet, John, Ragnarsdóttir Kristín Vala, 2010. "The History of The Limits to Growth" Solutions, Feb 26, http://www.thesolutionsjournal.com/node/569, (accessed Aug 10 2010)

Norgate, T., Rankin, J., 2002. "Tops at Recycling, Metals in sustainable development", CSIRO sustainability papers, http://www.bml.csiro.au/susnetnl/netwl30E.pdf (accessed Aug 31 2010)

O'Neill, Brendan, 2010. "Think the Earth is Finite, Think Again" http://www.spiked-online.com/index.php/site/article/9867/, (accessed Nov 19 2010)

O'Neill, Gerard, 1974. "The Colonization of Space" Physics Today, 27, September issue, 32–40

Odum, H.T., 1994. Ecological and General Systems: An Introduction to Systems Ecology, Colorado University Press, Colorado

Ong, E., Glantz, S.A., Hirayama's work has stood the test of time Public Health Classics, *Bulletin of the World Health Organization*, 2000, 78 (7), 938

Ordway, SH., 1953. "Resources and the American dream: including a theory of the limit of growth." Ronald, New York, p 281

Oreskes, N., and Conway, E.M., 2010. "Merchants of Doubt", Bloomsbury, New York.

Papert, S., 1980. Mindstorms; Basic Books: New York, NY

Passel, P., Roberts, M., Ross L., 1972. New York Times, April 2

Paul, G.S., and Cox, E.D., 1996. "Cyberevolution and future minds," Charles River Media, Rockland, MA

Pearce, Fred, 2010. "Peoplequake: Mass Migration, Ageing Nations and the Coming Population Crash" Transworld Publishers Ltd., ISBN 13: 9781905811342 ISBN 10: 1905811349

Peccei, A., and Ikeda, D., 1984. "Before it is too late", Kodansha International, Tokyo

Phillips, A.W., 1950. "Mechanical Models in Economic Dynamics", *Economica*, New Series, Vol. 17, No. 67, 283–299

Phillips, A.W., 1954. Stabilization policy in a closed economy, Economic Journal 64, p. 290-L321

Papp, J.F., 2005. "Recycling Metals", United States Geological report http://minerals.usgs.gov/minerals/pubs/commodity/recycle/recycmyb05.pdf (accessed Aug 31 2010)

Ponting, Clive, 2007. "A New Green history of the world" Vintage books, London

Popper, S.W. Lempert, R.J, Bankes S.C. 2005 "Shaping the Future", *Scientific American*, 292 (4):66–71

Prins, Gwyn, et al., 2010. "The Hartwell Paper – A new direction for climate policy after the crash of 2009". London School of Economics. http://eprints.lse.ac.uk/27939/1/HartwellPaper_English_version.pdf. (accessed Nov 14 2010)

Rábago, K.R., Lovins, A.B., Feiler, T.E., 2001. "Energy and Sustainable Development in the Mining and Minerals Industries", IIED report, http://www.iied.org/mmsd/mmsd_pdfs/041_rabago.pdf Retrieved Aug 31 2010

Rademaker, M., and Kooroshy, J., 2010. "The Global Challenge of Mineral Scarcity", Essay for the Conference: Enriching the Planet, Empowering Europe – Optimising the use of natural resources for a more sustainable economy, The Hague, 26 & 27 April 2010, http://www.clingendael.nl/resourcescarcity (accessed Sep 04 2010)

Radetzki, M.N., 2010. "Peak Oil and other threatening peaks—Chimeras without substance" Energy policy, in press

Ravaioli, C., and Ekins, P., 1995. "Economists and the Environment: What the Top Economists Say about the Environment". Zed Books, London, 212 pp

Reijnders, L., 2006. "Cleaner nanotechnology and hazard reduction of manufactured nanoparticles" *Journal of Cleaner Production*, Volume 14, Issue 2, Pages 124–133

Reynolds, D.B., 1999. The mineral economy: how prices and costs can falsely signal decreasing scarcity. *Ecological Economics* 31, 155

Reynolds, D.B., 2000. "The case for conserving oil resources: the fundamentals of supply and demand". OPEC Review (June)

Regis, E., 1994. "The Great Mambo Chicken and the Transhuman Condition". Addison Wesley, Reading, MA

Regis, E., 2004. "The Doomslayer" Wired Magazine, http://www.wired.com/wired/archive/5.02/ffsimon_pr.html, (accessed Nov 22 2010)

Revelle, R., and Suess, H., 1957. Carbon dioxide exchange between atmosphere and ocean and the question of an increase of atmospheric CO2 during past decades. Tellus 9: 18–27

Richardson, L.F., 1922. "Weather prediction by the numerical process", Cambridge University Press, London

Robock, A.L., Oman, and G.L., Stenchikov, 2007. Nuclear winter revisited with a modern climate model and current nuclear arsenals: Still catastrophic consequences, *J. Geophys. Res.*, 112, D13107, doi:10.1029/2006JD008235

Roma, P., 2003. "Resources of the sea floor" *Science* Vol. 299, no. 5607, pp. 673–674

Sadiq, M., 1992. "Toxic Metal Chemistry in Marine Environments" Taylor & Francis, Washington, ISBN: 0824786475

Sandback, Francis, 1978. "The rise and fall of the Limits to Growth debate", *Social Studies of Science*, Vol. 8, No. 4, pp. 495–520

Sandberg, A., Boström, N., 2008. "Whole Brain Emulation: A Roadmap. Technical Report #2008-3". Future of Humanity Institute, Oxford University. http://www.fhi.ox.ac.uk/Reports/2008-3.pdf. (accessed Sep 11 2010)

Schoijet, M., 1999. "Limits to Growth and the rise of Catastrophism" *Environmental History*, Vol. 4, No. 4, pp. 515–530

Senge, P.M., and Sterman, J.D., 1992. "Systems thinking and organizational learning: Acting locally and thinking globally in the organization of the future" *European Journal of Operational Research* Volume 59, Issue 1, Pages 137–150

Shen, H., Fossberg, E., 2003. "An overview of recovery of metals from slags" *Waste Management* Volume 23, Issue 10, Pages 933–949

Shoham Yohav, and Leyton-Brown, Kevin, 2008. "Multiagent Systems: Algorithmic, Game-Theoretic, and Logical Foundations", Cambridge University Press, Cambridge, ISBN 9780521899437

Shubik, M., 1971. "Modeling on a Grand Scale", *Science* vol 174, p 1014

Simmons, M., 2000. "Revisiting The Limits to Growth: Could The Club of Rome Have Been Correct, After All?" http://www.simmonsco-intl.com/files/172.pdf

Simon, J., 1981. "The Ultimate Resource" Princeton Unicersity press, Princeton

Skinner, Brian J., 1976. "A Second Iron Age Ahead?"*American Scientist*, vol. 64 (May-June issue), pp. 258–269.

Skinner, B., 1979. Proc. Natl. Acad. Sci. USA Vol. 76, No. 9, pp. 4212–4217

Slade, M., 1982. "Trends in Natural-Resource Commodity Prices: An Analysis of the Time Domain" *Journal of Environmental Economics and Management* vol 9, pp. 122–137

Smith, V.L., 1968. Economics of Production from Natural Resources. *Am. Econ. Rev.* 58, 409–431

Solow, R., 1957. The Review of Economics and Statistics, Vol. 39, No. 3, pp. 312–320

Solow, R., 1974. "The Economics of Resources or the Resources of Economics," *American Economics Review* 64: 1–14

Sohn, Ira., 2005. "Long-term projections of non-fuel minerals: We were wrong, but why?" *Resources Policy*, Volume 30, Issue 4, Pages 259–284

Spector, R., 2011. "The War on Cancer A Progress Report for Skeptics" The Skeptical Enquirer, Volume 34.1, January/February 2010. http://www.csicop.org/si/archive/category/volume_34.1, (accessed Feb 27 2011)

Staniford Stuart, 2010. "Patent Stat Charts" http://earlywarn.blogspot.com/2010/08/patent-stats-chart.html (accessed Sep 11 2010)

Starr, Steven, 2009. "Catastrophic Climatic Consequences of Nuclear Conflict" http://www.icnnd.org/Documents/Starr_Nuclear_Winter_Oct_09.pdf (accessed Nov 10 2010)

Stegerman, J.J., and Solow, A.R., 2002. "A Look Back at the London Smog of 1952" Environmental Health Perspectives Volume 110, Number 12, December issue. http://ehp.niehs.nih.gov/docs/2002/110-12/editorial.html (accessed Sep 03 2010)

Sterman, J.D., 2000. "Business Dynamics: Systems Thinking and Modelling for a Complex World", Irwin/McGraw-Hill: New York

Sterman, J.D., 2002. "All models are wrong: reflections on becoming a system scientist", *System Dynamics Review*, Vol 18, No. 4, pp. 501–531

Sterman J.D., and Booth L., Sweeney, 2007. Understanding Public Complacency About Climate Change: Adults' mental models of climate change violate conservation of matter. *Climatic Change*, 80, 213–238

Sterman, J.D., 2011. "A banquet of consequences" Climate interactive blog, http://climateinteractive. wordpress.com/2011/01/03/john-sterman-a-banquet-of-consequences-at-2010-sdm-confer- ence-at-mit/

Stiglitz, J.E., 1974a. Growth with Exhaustible Natural Resources: Efficient and Optimal Growth Paths. Review of Economic Studies, Symposium on the Economics of Exhaustible Resources, 123–37

Stiglitz, J.E., 1974b. Growth with Exhaustible Natural Resources: The Competitive Economy. Review of Economic Studies, Symposium on the Economics of Exhaustible Resources, 138–57

Strauss, L., 1954. "This Day in Quotes: SEPTEMBER 16 – Too cheap to meter: the great nuclear quote debate". This day in quotes. 2009. http://www.thisdayinquotes.com/2009/09/too-cheap-to- meter-nuclear-quote-debate.html. (accessed Aug 20 2010)

Sun Tzu, ca. 400 b.c.e. "The Art of War". In "Project Gutemberg" http://www.gutenberg.org/ files/132/132.txt accessed May 2010 – sec 25 of the 1st chapter 1.25

Tainter, J.A., 1988. The Collapse of Complex Societies. Cambridge: Cambridge University Press.

Tainter, J.A., 1996. "Complexity, problem solving and sustainable societiesy" in "Getting down to earth, Practical Applications of Ecological Economics", Island Press, Washington; ISBN 1-55963-503-7

Tarter, J., 2003. "Ongoing Debate over Cosmic Neighbors "Science, New Series", Vol. 299, No. 5603 pp. 46–47

Technology Review, 2010, "Europe's plan to simulate the entire planet" http://www.technologyre- view.com/blog/arxiv/25126/?ref=rss&a=f accessed 07 Nov 2010

Tierney, J., 1990. Betting the planet. New York Times Magazine December 2: 79–81

Tilton, J.E., 2006. "Depletion and the long-run availability of mineral commodities". In: M.E. Doggett and J.R. Perry, Editors, Wealth Creation in Minerals Industry: Integrating Science, Business and Education: Special Publication 12, Society of Economic Geologists, Littleton, CO.

Tilverberg, Gail, 2011. "There is no steady state growth, except at a veri basic level" http://ourfi- niteworld.com/2011/02/21/there-is-no-steady-state-economy-except-at-a-very-basic-level/ (accessde 21 Feb 2011)

Tinbergen, J., 1977. "RIO, Reshaping the International Order, a Report to the Club of Rome". New American Library, New York

Townsend, Robert, L., 1970. Up the Organization; How to Stop the Corporation from Stifling People and Strangling. New York: Alfred A. Knopf

Turco, R.P., Toon, O.B., Ackermann, T.P., Pollack, J.B., and Carl Sagan, 1983. "Nuclear Winter: Global consequences of multiple nuclear explosions", Science, Vol. 222, No. 4630, December issue, pp. 1283–1292

Turco, R.P., 2007. Climatic consequences of regional nuclear conflicts, Atmos. Chem. Phys., 7, 2003–2012

Turner, G.M., 2008. "A comparison of The Limits to Growth with 30 years of reality" Global Environmental Change 18 (2008) 397–411

Tustin, A., 1954. "The Mechanism of Economic Systems: An approach to the problem of eco- nomic stabilization from the point of view of control system engineering" The Economic Journal, Vol. 64, No. 256, pp. 805–807

Tyndall, J., 1863. "On radiation through the earth's atmosphere", Phil mag, 25, fourth series

United Nations, 1987. "Report of the World Commission on Environment and Development." General Assembly Resolution 42/187, 11 December 1987

Vajk, P.J., 1976. "The Impact of Space Colonization Upon World Dynamics." Technol. Forecast. Soc. Change 9, pp. 361–400

Vajk, P., 2002. http://settlement.arc.nasa.gov/CoEvolutionBook/LIFES.HTML#Limits%20to%20 Growth-wronger%20than%20ever (accessed 28 Aug 2010)

Valero, A., Valero, A., 2010. "A prediction of the energy loss of the world's mineral reserves in the 21st century", Energy, In Press

Verheijen, F.G.A., Jones, R.J.A., Rickson, R.J., Smith, C.J. 2009. "Tolerable versus actual soil erosion rates in Europe", Earth Sciences Review, Volume 94, Issues 1–4, Pages 23–38

Victor, P.A., 2008. "Managing without growth: slower by design, not disaster" Northampton, MA, Edward Elgar

Vinge, Vernor, 1993. "The Coming Technological Singularity" http://www-rohan.sdsu.edu/faculty/vinge/misc/singularity.html

Volterra, V., 1926. "Variazioni e fluttuazioni del numero d'individui in specie animali conviventi". Mem. R. Accad. Naz. dei Lincei. Ser. VI, 2, 31–113

Ward, P., 2009. "The Medea Hypothesis: Is Life on Earth Ultimately Self-Destructive?" Princeton University Press, Princeton, ISBN13: 978-0-691-13075-0

Weart, S.R., 2003. "The Discovery of Global Warming", Cambridge, Harvard Press

Warren Thompson, 2003. "Encyclopedia of Population". MacMillan Reference, pp. 939–940. ISBN 0-02-865677-6

White, Steven, 2002. "Rant: we love this stuff" http://www.stevewhite.org/log/archive/20020707.htm (accessed Aug 27 2010)

Wilensky, U., 1999. NetLogo. http://ccl.northwestern.edu/netlogo. Center for Connected Learning and Computer-Based Modeling, Northwestern University, Evanston, IL

Wikipedia, 2010. "Abundance of the elements (data page)" http://en.wikipedia.org/wiki/Abundances_of_the_elements_%28data_page%29 (retreived 31 Aug 2010)

Zimmermann, Erich, 1933. World Resources and Industries. New York: Harper & Brothers. 1951. World Resources and Industries, 2nd revised ed. New York: Harper & Brothers

Index

Made in the USA
Middletown, DE
29 July 2021